658.562 DOB

BOOK NO: 1864235

D0303507

MAY BE REI
UNLESS R
ANO
PHONE (
FINE 1

ISO 9001:2000 Quality Registration Step by Step

UNIVERSITY OF WALES, NEWPORT
LIBRARY
AND
INFORMATION
SERVICES
ALLT-YR-YN

MAY BE REI
UNLESS R
ANO
PHONE (
FINE 1

UNIVERSITY OF WALES, NEWPORT
LIBRARY
AND
INFORMATION
SERVICES
ALLT-YR-YN

ISO 9001:2000 Quality Registration Step by Step

Third edition

UNIVERSITY OF WALES, NEWPORT
LIBRARY AND INFORMATION SERVICES
ALLT-YR-YN

F. P. Dobb BSc CEng FIMechE FIQA RSC
QMS 2000 Principal Auditor

ELSEVIER
BUTTERWORTH
HEINEMANN

AMSTERDAM BOSTON HEIDELBERG LONDON NEW YORK OXFORD
PARIS SAN DIEGO SAN FRANCISCO SINGAPORE SYDNEY TOKYO

Elsevier Butterworth-Heinemann
Linacre House, Jordan Hill, Oxford OX2 8DP
200 Wheeler Road, Burlington MA 01803

First published 1996
Second edition 1999
Third edition 2004

Copyright © 2004, F. P. Dobb. All rights reserved

The right of F. P. Dobb to be identified as the author of this work
has been asserted in accordance with the Copyright, Designs and
Patents Act 1988

No part of this publication may be reproduced in any material form (including
photocopying or storing in any medium by electronic means and whether
or not transiently or incidentally to some other use of this publication) without
the written permission of the copyright holder except in accordance with the
provisions of the Copyright, Designs and Patents Act 1988 or under the terms of
a licence issued by the Copyright Licensing Agency Ltd, 90 Tottenham Court Road,
London, England W1T 4LP. Applications for the copyright holder's written
permission to reproduce any part of this publication should be addressed
to the publisher

Permissions may be sought directly from Elsevier's Science and Technology
Rights Department in Oxford, UK: phone: (+44) (0) 1865 843830;
fax: (+44) (0) 1865 853333; e-mail: permissions@elsevier.co.uk. You may also
complete your request on-line via the Elsevier Science homepage
(www.elsevier.com), by selecting 'Customer Support' and then
'Obtaining Permissions'.

British Library Cataloguing in Publication Data
A catalogue record for this book is available from the British Library

Library of Congress Cataloguing in Publication Data
A catalogue record for this book is available from the Library of Congress

ISBN 0 7506 4949 6

For information on all Elsevier Butterworth-Heinemann publications
visit our website at: www.bh.com

Composition by Genesis Typesetting Limited, Rochester, Kent
Printed and bound in Italy

Contents

Preface

With its model Quality Policy Manual and model Operating Procedures, this book is an unashamedly do-it-yourself, step-by-step manual or workbook. Using this book, any small to medium sized enterprise can, on its own, achieve registration to the ISO 9001:2000 standard.

The idea for this book was born whilst I was manning an exhibition stand at the NEC in Birmingham for my then employers, National Quality Assurance. During those two days I personally had five different managing directors of small firms come and ask me for a model of what was required to meet ISO 9000. This has been a constant and consistent request or demand of potential registrants ever since.

Also, at that time, I assessed two companies who had documented and implemented exemplary management systems, with little outside assistance. One was a small firm, in west Wales with a staff of two, which installed insulation materials; the implementation had been done jointly by a father (one of the partners) and his daughter as part of a university project. The other was a cleaning contracting company employing about 200 in the west of England, where the financial accountant had read the standard, looked at one or two examples in various companies and then set about documenting and implementing one of the most professionally produced and simple systems I have ever seen.

It is stated by many, including myself, that your documented management system is unique to the way in which you run your business, and cannot be taken off-the-shelf. However, what is hardly ever stated, but is equally true, is that the **best method or format of presenting management systems has now become almost uniform.**

This book, with its step-by-step approach and its model quality manual and model operating procedures, provides a workbook and the template around which you need to build your system. The model operating procedures include alternatives for the majority of industries and management systems; hence, do please note your final operating procedures should have **substantially fewer pages than the models provided.**

Despite extensive experience and adequate qualifications, I have never had the courage to start my own business and I have a considerable admiration for those individuals that have. I dedicate this book to such entrepreneurial individuals and trust this book will not only enable them to achieve ISO 9001 registration but provide a basis for an effective, efficient and economic documented management system that allows them to survive and prosper.

Fred Dobb
Garndiffaith, Wales, U.K.
fred@peterdobb.freeserve.co.uk

Preface

Preface to third edition of ISO 9001:2000

I was requested by the publishers to amend this book for publication at the begining of 2001 to coincide with the publication of the revised standard. I resisted the tempation to be one of the first off the block. I was concerned that with some new requirements, it was not at all clear how they would be interpreted or addressed. With benefit of hindsight, for once, I'm glad I made that decision. The initial variation has been unbelievable. After reviewing the situation and also assessing many efforts, I am pleased to provide a pragmatic solution.

For the academics among the readers of my text, I apologise for the simple English and use of words. From the letters I received, it was clear that many readers', or customers', first language was not English.

Fred Dobb

Free web downloads

Electronic versions of the templates in the book are available on the Web at http://books.elsevier.com/manuals/0750649496. They include:

- Draft procedures
- Draft Quality Policy Manual
- Templates or drafts of typical forms

This material is only available to purchasers of this book. To obtain a password please e-mail Jo Blackford at j.blackford@elsevier.com or telephone her on +44 1865 314447 making sure you have a copy of the book in front of you.

For up-to-date information on all Elsevier Quality improvement books visit our web site at http://books.elsevier.com/industrialeng. You can also register to receive regular e-mail bulletins.

Acknowledgements

I would gratefully acknowledge my wife and boys for their patience whilst I undertook this project. I must admit it was far more demanding in time and effort than I appreciated.

I would also acknowledge the dedicated skills of my wife who checked the English and spelling, and Mrs Sandra Brice, MIQA, who typed (and retyped) the script and also constructively criticised the input, at appropriate stages.

About the author

The author was very fortunate in being able to complete a full five-year industrial engineering apprenticeship at the famous Plant Locomotive Works in Doncaster, Yorkshire before pursuing a full-time degree course at the University of Wales, Cardiff.

This has given him a very pragmatic and practical approach to management based on plain common sense. In his opinion, many of the faults of UK industry is that at intermediate and senior management grades, common sense is not always prevalent. He is often in despair about the lack of basic practical engineering knowledge and skills gained by engineering graduates but is greatly encouraged by the sudden and spectacular return of formal apprenticeships into UK industry since 1995.

After graduating in 1966, his early career was in plant engineering in chemical plants working for Monsanto, British Aluminium, INCO and Dunlop. In 1974 he was recruited as a Plant Engineer by the Royal Ordnance Factories. Prior to actually commencing his employment with the ROF a new priority was outlined, namely to meet the new Quality Assurance Standard Def. Stan. 05/21 (AQAP–1). Fred was actually informed of his new responsibilities at 9.15 a.m. on his first day and made successful career progression within the Royal Ordnance Factories for eleven years, specialising for the first seven years on QA aspects of explosives and ammunition, followed by senior management of design and then manufacture.

Following the privatisation of the UK ordnance factories, it was with great regret that Fred decided to leave this employment to become QA Manager for the world's leading diaphragm valve manufacturer, Saunders Valve, followed by the position of founder executive director of The Wales Quality Centre. He was then given the task of creating a QA consultancy for WS Atkins Consultancy, Wales and the South West. This consultancy was highly successful with over 60 firms achieving registration to ISO 9001/2 at their first attempt.

Fred has been a Registered Lead Assessor since 1989.

He has worked as senior manager–principal assessor in several leading certification bodies. He is now the Commercial Manager covering UK and also Eastern Europe for one of the leading Certification Bodies, United Registrar of Systems (or 'URS').

Chapter 1
Clear priorities

Before you start the project to obtain ISO 9001 registration, it is **essential** that you realise there are **two** clear priorities. **Both** must be achieved to reap the full benefit of ISO 9001 registration.

The first priority

It is essential that the management system and procedures that you introduce are:

- effective
- efficient
- economical
- enable you to continually improve.

I cannot stress **economical** enough. You will undoubtedly have seen, in the press, examples of where it is claimed it cost tens of thousands of pounds/dollars to introduce ISO 9000 in a small firm of 10, where it is now necessary to have forms and procedures for this, that and everything, or where it slows responses to customers, etc., etc.

> Quite simply, **people who believe that it is necessary to have such over-documented systems in order to achieve registration have been totally and completely misled.**

If you do not get this message clear from the start, and you obtain registration with an over-documented system, the only benefit you will achieve is an increased order book without financial gain.

It is not uncommon for the author, as a Lead Assessor, to tell the chief executive of a company that I am delighted to recommend his/her organisation for registration, while at the same time in truth being really very sad. The system I have assessed, and which meets the objective criteria set, is often completely over-documented and is going to cost this organisation a fortune to run.

It will become clear how this can easily arise when you read 'Advice on consultants' and 'Plan for and stage manage the assessment'.

The second priority

The next priority is to obtain ISO 9001 registration as painlessly as possible, in a sensible time-scale and at a low overall cost.

This is the end of the first chapter. It may be short, but if you've got the message about the dual priorities, you're better educated and more knowledgeable than fifty per cent of the world's decision makers, academics, 'consultants' and technical press.

Chapter 2
Why bother?

There is no one clear-cut reason for seeking registration to ISO 9001. The reasons are many and varied. The more common reasons given are:

- Mandatory requirement of customers;
- Reduction of multiple assessments by different customers;
- Necessary to survive;
- It is good business practice;
- It gives management control;
- To cut the cost of poor quality;
- Protection against criminal prosecution and civil claims;
- For marketing advantage.

Mandatory requirement of customer

Many organisations are now making it mandatory that subcontractors or suppliers that provide them with services and products must be registered to the International Standard ISO 9001. This is quite fair and sensible in both principle and practice, as long as it is not abused or overstated. The argument for insistence upon ISO 9001 registration is as follows.

Large corporate buyers buy from hundreds if not thousands of subcontractors and suppliers. Often these purchases will be made on a basis of competitive tender. For the large corporate buyer to visit each and every one of these suppliers (or potential suppliers) and carry out a full and comprehensive audit would be prohibitively expensive. Also a large team of experienced assessors would have to be recruited and there would not be enough experienced and qualified assessors to go around.

Hence from this point of view it makes much more economic sense for these buyers to have their suppliers (and potential suppliers) systematically and independently assessed and audited by experienced and qualified assessors. Assessment being carried out to known standards and procedures by certification or registration bodies that have themselves been thoroughly assessed and constantly audited by a government-approved body (e.g. accredited by UKAS the United Kingdom Accreditation Service, previously known as the NACCB; other countries may have similar approval bodies).

Note *You may be offered certification or registration by an individual who claims to be an 'independent assessor', or by companies (often 'phoney' companies with fancy names covering sole traders or partnerships) offering you cheap consulting and at the same time registration to ISO 9001. Do not be fooled. Firstly, the certificate they will give you may be worthless and will not stand even cursory examination by a knowledgeable buyer. Secondly, are they really that cheap – is an unaccredited certificate worth anything? If you check the fee structure of the accredited certification bodies , see appendix A, you will generally find them very reasonable. They are now operating in a **very** competitive market. Also in UK you may wish to check if the accreditation is by the government approved body UKAS, with the accompanying 'Tick & Crown' logo. Unbelievably enterprising individuals have even set up an unofficial accreditation body to provide 'accreditation' for such bodies.*

For large corporate buyers to have all their suppliers (or potential suppliers) assessed in this way provides a level playing field, with everybody working to at least the same minimum quality assurance standard.

Insistence of supplier conformance to ISO 9001 also cuts out the 'cowboys'. To have passed initial assessment and continuing surveillance to achieve registration, even the smallest organisation must exist in fact as well as on paper, and must be consistently producing the product or service offered to known and/or contractually agreed standards.

This sensible and logical approach has been cascaded down and applied in smaller organisations. However, it has in some cases been taken to silly extremes. Examples of abuses are:

- One very large UK utility notified all its suppliers by letter that, as of the following month, business would only be placed with ISO 9000-registered organisations. One firm the author was working with had to immediately reduce its workforce from 36 to 17 due to lack of orders. This was totally unreasonable; at least one year's notice should have been given.
- One world-wide oil company insisted that the mother and son team, providing and watering potted plants in the corporate headquarters, become ISO 9000 registered.
- 'Supplier questionnaires' can proliferate and arrive like junk mail. Such questionnaires have their legitimate uses, but it is **not** an essential requirement of ISO 9001 that questionnaires be sent to every supplier. If you have a consultant who tells you it is essential, perhaps you ought to question his motives – especially if he offers to send them and analyse the replies for you.
- A supplier questionnaire was sent out by a prestigious UK City Council to the squire of the local Morris Dance Team. (Morris men are traditional folk dancers, who usually perform outside rural public houses). A follow-up letter arrived a few weeks later, threatening to take them off the supplier list if they did not complete and return the questionnaire (I'm sure an enterprising QA consultant would have sorted them out. Written procedures for calibrating sticks and hankies for size, colour and cleanliness. Assessment of the hostelries they dance outside. Written instructions for 'Ye dance of Young Harry from ye village of Olde Thistlebottom in the County of Somerset'. It sounds like it could be good fun. I'd do the consultancy for free!)

Unfortunately the above abuses and silliness attract the press and gain adverse publicity.

The Certification Bodies and UKAS do try and spread the gospel of good sense in using registered suppliers, but unfortunately it does not attract the same press interest or coverage as conflict, distress, complaints or sheer stupidity.

When looked at rationally and logically, the case for insisting upon ISO 9001 registration by suppliers of quality critical items is very sound.

Reduction of multiple assessments by different customers

Unfortunately, people have short memories. Before ISO 9001 (or BS 5750 as it was) came into existence, as a QA manager in large manufacturing organisations I had to deal with at least one visit or audit from potential customers every single week.

These could be very time-absorbing and often wasteful. In a large percentage of the cases, it was a case of a brief site tour to impress them with the computer-controlled machining centre and the computerised SPC, before banging off some coloured grenades on the proof yard and heading off to lunch. If the visitor came from abroad possibly a shopping trip to Cardiff had to be included or a visit to the 12th-century castle at Caerphilly etc. These, often unnecessary, visits were common throughout British industry in the 1960s and 1970s.

After ISO 9000 registration, a visit from out customers or formal audits became much rarer. Often when such a visit was proposed, a copy of our ISO 9000 certificate plus possibly a copy of the Quality Manual was enough to satisfy them and persuade them that their visit was not necessary to confirm our suitability. When you are registered you should be able to do the same.

The ability to survive

As I will show later, ISO 9001 is simply a case of good management systems and practice with quality assurance built in. This ensures consistency of product or service.

If your organisation supplies inconsistent, poor quality or completely wrong products or services, your company or its products will not survive, or your reputation will become heavily tarnished.

Sometimes it can be very quick and very dramatic:

- In recent years one very large producer of health products was found to have contaminated a very small percentage of its baby foods, reported as due to an infection in its automatic cleaning processes in the factory. These food containers were returned by mothers and cleared from the shelves of shops. A company with a turnover of over £30 million a year went into liquidation in a couple of weeks.
- Another incident in the not-too-distant past involved tinned salmon. It was reported that botulism in tinned salmon had killed 3 or 4 pensioners. The price of **every** manufacturers' tinned salmon dropped immediately and completely and took about ten years for the trade to recover.
- In 1990 it was reported that one manufacturer of mineral water had contaminated its product with benzene and had to withdraw 160 million bottles from supermarket shelves world-wide. A similar incident was reported in 1995 with a different manufacturer of mineral water contaminated with dichlorobenzene, leading to a withdrawal of 3 million bottles of a self-branded product from the shelves of a leading UK supermarket.
- In a remote part of Wales, a farmer tried to get a telephone installed and was told it would cost many thousands of pounds. Subsequently the phone company advertised a special offer whereby they would install any new phone for £99. It was reported that the farmer's enterprising wife filled the form in and sent it off, and it was accepted. On realising their mistake, the phone company tried to renege on their contract but ultimately they had to install the phone for £99 as that was the contract they had accepted. The phone company could stand such a loss, could you?

Good business practice

When you've read further on, you will realise that ISO 9001 should **not** be about producing documents and numbering them in a certain way to satisfy some pedantic assessor.

ISO 9001 (when de-jargonised) is straightforward good business practice. All it is really saying is, before you accept a contract or send a tender ensure:

- you have all the necessary information and you effectively communicate with your customers;
- you are sure you can actually supply the product or service;
- that what you propose to supply meets legal and regulatory requirements;
- you actually want to do it for the price and conditions offered.

Having got the contract, you buy any items, materials or services required:

- clearly specifying in writing what you want
- only buying from suppliers who you have evaluated and consistently supply good products or services;
- when goods arrive or services are provided you will verify that they are correct before use.

You communicate with your staff to ensure they are aware of the importance of their input and have been trained and are competent for the tasks they are asked to do. If staff need written procedures or instructions, you provide them (in the same way that at home you would automatically use a knitting pattern for a jumper, a recipe for a cake or use a workshop manual for your car).

You routinely analyse what you are doing and the results and set quality objectives and strive for continual improvement.

If you look through the clauses of the ISO 9001 Standard and really understand them, you will find that they are just plain common sense, excellent business practice and things you should be doing anyway.

Furthermore, these controls with their in-built feedback give a formal overview of management decisions and actions and company operating efficiency. By using this information and the controls that ISO 9001 makes available, it is possible to get substantial improvements in business performance – and, just as importantly, to maintain those improvements.

Management controls

When you have good documented systems in place, you will have simple and effective controls. These will then ensure that, as far as is possible, your company is not dependent on any one person. Hence your company is less likely to fail because a key individual is on long-term sick, has died or has left the company. You may even consider going on holiday yourself without wondering if the company will falter.

It also provides the basis for controlled and managed growth. It is unlikely that a firm will ever make the transition from small to medium size without documented management systems. There is a limit to the number of staff or projects you can manage by writing notes on the back of a cigarette packet!

When you have your documented management system in place, you will have time for new projects, development of new markets, etc. The idea of having simple, effective documented systems is that the everyday routine becomes just that – routine. As much as is possible of the business ticks over and functions just as naturally as your left foot follows your right.

The other management control or discipline that ISO 9001 introduces is consistency of output. As chief executive, you do not want variation in quality because of poor raw materials, processes or pressure to meet end-of-month targets, etc. These invariably lead to customer complaints. Fully implementing a simple and effective management system based on ISO 9001 removes these uncertainties. It will reduce the problems of the unexpected, unwanted and often fully justified customer complaints. Nobody likes going on the phone to eat 'humble pie'. (I remember I once flew all the way from Heathrow to Washington just to allow a customer to vent his wrath on me because of the deficiencies of the company. Nobody wants or needs this unless you are a masochist.)

Other controls and improvements will come via an effective internal audit system carried out by those independent of the functions being audited. If carried out effectively, these audits can become the eyes and ears of the Managing Director, cutting right across the barriers that would otherwise filter out the information the managers don't want the managing director to hear.

Another control that ISO 9001 can give, one that is difficult to achieve by any other means, is consistency of operation between different groups, sites or offices in the company, irrespective of who is managing them. Implementing ISO 9001 and going for certification enforces consistency across the organisation.

Cutting the cost of poor quality

The considerable costs that poor quality inflicts on an organisation are rarely recognised and are hardly ever quantified in terms of pounds or dollars lost directly off the bottom line.

By introducing simple and effective management and quality assurance controls based on the principle of ISO 9001, the cost of poor quality can be reduced significantly. In almost every case where the project has been carried out with the priorities firmly in mind (see Chapter 1), the savings made year-on-year will more than pay for the costs of introduction.

The cost of poor quality

It is now well established among practising quality professionals that there are four separate elements to quality costs.

Two elements contributing to the cost of poor quality:

- The cost of external failures
- The cost of internal failures.

Two elements arising from money spent to reduce the costs of poor quality:

- The investment in appraisal (inspection)
- The investment in prevention (quality assurance systems).

The cost of **external** failures arises directly from replacing or refunding customer rejects, and from customer warranty support. Also there are huge indirect costs of handling customer complaints and rejects/returns: meetings, design changes, liability claims, letters, more meetings, special visits, business lunches, special transport etc. The greatest cost of external failure can be the indirect cost due to loss of future business because a client moves his order elsewhere, or even worse tells others how bad you are, i.e. you lose your reputation. These indirect costs of external failures are hidden and rarely quantified. They will usually far exceed the direct costs.

The cost of **internal** failures arise directly from:

- scrap and replacement costs of raw materials and bought-in items;
- double labour costs for rework or repair;
- inspection and re-inspection costs;
- damage to machine tools and plant;
- injuries to staff.

Again there are huge indirect costs: administrative and supervisory time, frustration and demoralisation, production schedules disrupted, failure investigations, late deliveries etc. And once again, these huge indirect costs are rarely quantified.

The **investment in appraisal** is the direct cost of inspection time and equipment including calibration and maintenance of inspection equipment. The inspection may be on materials, components, work-in-progress and final product. It may also include some product verification or product audit.

The **investment in prevention** is the direct cost of quality engineering and planning, the cost of developing and implementing your quality system. The indirect costs may include more time spent on product (or service) design, better training and education of staff, more time or money spent on equipment procurement and maintenance. There may also be a reduction of throughputs to allow for more attention to the quality of the service provided or goods manufactured.

What must be achieved is a balance between the four elements, in order to reduce the overall cost of quality (see Fig. 2.1).

Note: Do be aware that, in some cases, there may be an initial jump in internal quality costs (see Fig 2.1) if staff have not been working to specification and are suddenly told to conform.

However, experience clearly demonstrates that investment in prevention and the effective introduction of ISO 9000 immediately reduces the cost of external failures and soon has an impact on the cost of internal failures. The net result is a substantial reduction in total quality costs.

Protection against criminal prosecution and civil claims

In certain industries (e.g. electrical manufacturing, pharmaceutical) the introduction of documented management systems with in-built quality assurance has successfully pre-empted, or has been offered as a defence against civil claims and criminal prosecution (e.g. involving the *Consumer Protection Act*, *Food Safety Act*, etc.).

Effective documented procedures (especially when independently certified by an accredited body) can be used to demonstrate that due diligence and reasonable care and precautions have been taken. The chief executive can justly claim that formal systems have been introduced which meet the International Standard ISO 9000 and which have also been formally and independently assessed by an accredited certification body. Furthermore they are regularly assessed by that body on an ongoing basis. The existence of proper documentary evidence to support your case is now more important than ever; under recent legislation the burden of proof generally falls upon you, the defendant.

In some sectors of industry, individual companies have significantly reduced their insurance policy costs in areas such as product liability and professional indemnity as a direct result of accredited registration to ISO 9000.

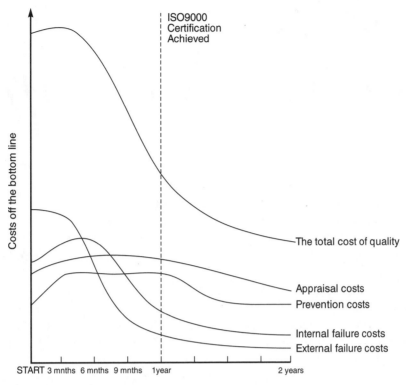

Fig. 2.1 Quality costs

Marketing advantage

In many sectors of industry, there is considerable status and marketing advantage to be had if your organisation holds independent quality assurance approved to ISO 9001.

It is almost as if the organisation gaining certification had graduated with honours. Even when the largest of organisations achieves registration, the chief executive is usually very proud to arrange a formal certificate presentation with the appropriate photographs for local and trade press.

For small organisations in particular, there is an enormous marketing gain. Prior to registration, the large or multinational competitor usually managed to dismiss the smaller firm as 'cowboys', 'not dependable', 'lacking expertise', 'insufficient resources', etc. When the smaller firm has achieved the accolade of an accredited third-party certification, they can not be dismissed so lightly. Salesmen from the newly registered small company find that buyers now give them appointments and treat them with due respect as a reputable and proven supplier.

It should also be noted that on accredited registration the firm should be listed in the United Kingdom DTI QA Register.

Chapter 3
The basic principles

Quality assured, certified, registered or accredited?

The above terms are very often confused and used incorrectly.

The firm which has been successfully assessed is either 'registered' or 'certified'. These terms are synonymous and the terminology is used by different certification bodies to mean exactly the same.

The firm or organisation is assessed by a Registered Lead Assessor working for an **accredited certification body**. This means that the certification body itself has been assessed and found to meet the standard required, and hence the certification body is 'accredited' by the appropriate government department. In the UK this is UKAS (previously the NACCB), representing the Department of Trade and Industry.

Note that the accreditation given to a certification body is not completely general in its scope. UKAS carefully considers the experience of the certification body and its assessors, and only gives accredited approval in those particular sectors of industry or commerce where adequate experience can be demonstrated.

You will hear people say that a firm has been accredited to ISO 9001/2. This is incorrect. The firm has been either 'certified' or 'registered'.

Where did ISO 9001 come from?

Figure 3.1 portrays a simplified evolution of quality assurance standards. The ISO 9000 family of standards arose following the very successful application of quality assurance standards throughout the British Ministry of Defence and its large procurement agencies during the 1970s and 1980s. The Ministry of Defence Procurement Executive was huge, even as late as 1989 MoD (PE) was the largest single customer of UK industry with £8.2 billion spend on equipment and a £3.4 billion spend on works, stores and services.

Prior to the implementation in 1973 of QA standards, the Ministry of Defence (MoD) employed literally an army of 16,500 civilian and military MoD inspectors. At that time there were at least ten QA departments with separate inspectors for each branch of the services (Army, Navy, Air Force and Ministry of Technology) and even specialist inspectors within each service (QAD(W), INO, DGShips, etc., etc.). A firm supplying the MoD would produce goods for the Ministry and then call in the MoD inspectors from the appropriate department when the goods were completed or at defined stages during manufacture. It was also common for the MoD inspectors to insist on inspecting raw materials, castings, forgings, timber etc. at source. Unannounced visits by these inspectors were part of the order of the day. In this situation inspectors' work could and did expand to occupy the number of staff and the time available.

In the early 1970s, following the Raby and the Raynor reports (of 1969 and 1971 respectively), the existing AQAPs (Allied Quality Assurance Publications) were adopted and introduced, principally as a contractual device to reduce the size of this visiting army of MoD inspectors. These standards took the responsibility of achieving the required specification away from MoD inspectors and put the onus upon the suppliers themselves to introduce effective Quality Assurance and Control. AQAPs were effectively introduced in 1973 and 1974 and retitled for a time as UK Ministry of Defence Def. Stan. 05/21, 05/24 etc.

It must be said (and the author was the factory chief inspector of a large ammunition factory employing over 2000 staff at the time) that, after the initial teething troubles, the introduction of these QA standards into the defense procurement industry was an overwhelming success. The vast army of inspectors reduced steadily to an appropriate size to enable the MoD to vet contracts and complete additional verification, when and if appropriate.

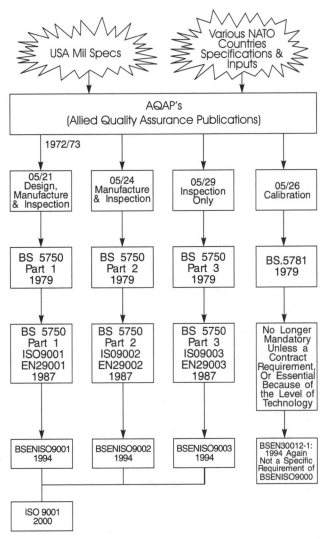

Fig. 3.1 Evolution of quality assurance standards

I must admit, this was not what I and many others expected at the time. I was very sceptical. I was terrified that when our own independent direct inspection of goods before dispatch from the suppliers was taken away, we would automatically be supplied a lower quality and many batches of non-conforming material. I was completely wrong. The reverse of my fears happened. As the firms were now responsible for and had a direct interest in the quality of their products, their controls and processes improved significantly.

Pressure therefore arose from private industry for the British Standards Institution to introduce an equivalent standard for the private sector of industry. Hence the Quality Assurance System Standard BS 5750:1979 was born. These were amended in 1987 when they were adopted by the International Organisation for Standardisation (ISO), as the ISO 9000 series. They were also accepted at that time by the European Committee for Standardisation (CEN – Comité Européan de Normalisation) as the EN 29000 series of standards. The ISO 9000 series of standards was further amended in years 1994 and 2000. Hence the standard usually designated as ISO9001:2000. They are now accepted and recognised in virtually every manufacturing and/or trading nation throughout the world.

You may see 'Guidelines for the interpretation of ISO 9000 for .... industries or business sectors'. I would recommend that you do **not** read these at the start of your ISO 9000 project. Many have been written by the good, the great and the academics for that industrial sector. These are therefore often misleading, because the authors may have rarely, if ever, documented and successfully implemented ISO 9001 in practice. They are usually based on a learned committee's considered reflections (and mutual compromises) on what the standard ought to mean for their particular industry written by representatives who may not have had 'hands-on' experience.

Therefore my advice is to purchase **only** ISO 9001:2000. With the practical guidance given in this book you can interpret the standard as it applies to your **own** particular business and your **own** management systems.

This may be purchased from:

Stationery Office, London
Tel. 0207 2426410 (also 0870 600 5522)   Fax 0207 2426410
www.clicktso.com

Or from,

I.L.I.,
Index House,
St. Georges Lane,
Ascot, Berks., SL5 7EU.
Tel. 01344 874343   Fax 01344 291194.

What ISO 9000 is really about

When a new factory or organisation is created without experienced advice it will almost invariably have developed and introduced, by intuition, a management system or process along the lines shown in Fig. 3.2. This is mainly based on the incorrect principle that Quality can be inspected into a product or service. Or, even more likely that the Quality of product (or introducing inspection) was an afterthought.

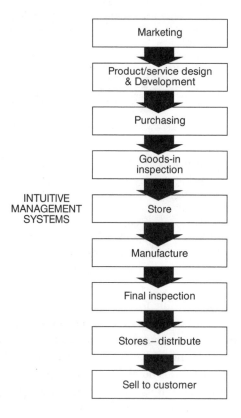

Fig. 3.2 Intuitive management systems

This may be fine initially on a greenfield site with an established and fully developed product. However, as output increases or new products are introduced, the problems that will arise are shown in Fig. 3.3.

It is obvious that a system which relies solely on inspection to provide a quality product or service is going to fail because the defects are found too late, i.e. when the material goods or services are already made or supplied. The defective goods or services must then be reworked, rectified, replaced, or the customer refunded. This is very expensive.

The other reason why it will fail is because people mistakenly believe that inspection of every individual product or item made (usually called '100% inspection') will provide defect-free goods or services, see Fig. 3.4.

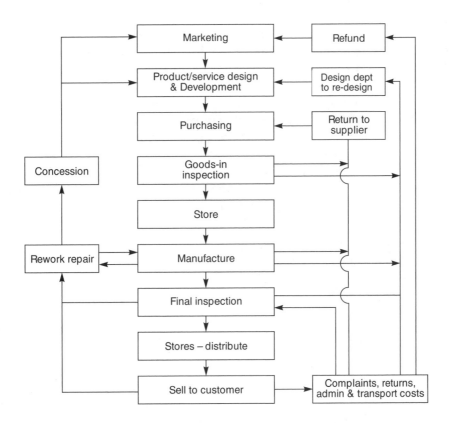

Fig. 3.3 Failures of intuitive management systems

It is a common misconception that the the provision of 100% inspecton will guarentee 100% perfect goods.

Fig. 3.4 A fallacy

People know in everyday life that 100% inspection does not work, e.g. after you have weeded the flower bed you will invariably find the odd weed which you have missed. How many times do you count your change in your purse before you decide what sum is right? Get 10 people to count the number of 'a's on this page: will they all record the same number? And so on.

Yet you often hear in work, 'This is a critical item or feature so we will 100% inspect it'. Always remember 100% inspection does not work! Look at Fig. 3.4 again: I trust you spotted all three mistakes! No? Perhaps you were not motivated or too tired to read it or perhaps your attention was diverted elsewhere. This is what happens in real life. Being an inspector tends to be a tedious and boring job.

The effectiveness of critical 100% inspection is shown in Fig. 3.5. You will miss more of the defective items as the job becomes more complicated, or there is a lower percentage of defects to find, less time is available, increased noise, repeated distractions, etc.

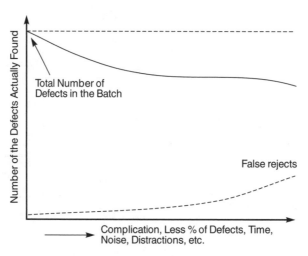

Fig. 3.5 Reduction in effectiveness of 100% inspection.

Both missed defects and 'false rejects' (e.g. items that *are* to specification that have been erroneously rejected by the inspector) are costly, particularly if re-inspection is demanded. I include two actual examples of failures of 100% inspection, which may surprise you and are thought provoking.

Case histories

A true tale of the inspector reacting to peer pressure
We were 100% inspecting some critical containers that were made from clear plastic. The process was not fully developed but the product was in manufacture due to urgent need and these containers had to be gauged and visually inspected for size and numerous subjective visual defects judged to be below specification. Typical reject rates in batches of 2000 containers at that time were between 10 to 20% with an experienced inspector taking 2 hours to inspect a batch. One day three experienced inspectors were checking these items with the above typical reject rates. Inadvertently, one of the inspectors was given a batch to inspect that had already been 100% inspected and the rejects removed. What percentage do you think was rejected out of this already cleansed' batch? After three-and-a-half hours (i.e. nearly twice as long as normal), another 12% were rejected! End of true story.

A true tale of an inspector on 'automatic pilot' or day dreaming
An inspector was inspecting small filters for penetrations (i.e. holes) on a salt particle test. We had automated the machine so that if the filter was good, a light in the console showed green and if defective it showed red, each test taking approximately five seconds. I was watching this inspector with no particular interest from the other side of the shop and noting and congratulating myself on how smoothly and efficiently we had developed the test, e.g. filter in, cover down, press foot pedal to start the test, wait approximately five seconds for the test cycle to complete, green light, cover up, remove accepted filter in right hand, insert next filter with left hand, cover down, press foot pedal, inspector stamps personal workmark on

the last filter to be tested (still in right hand), place in accepted tray, green light on, etc. etc.

This inspector had been testing filters for probably two hours without a significant break and was chatting amiably to her neighbour. She then had a red light for a reject. I watched with slightly more interest and then with horror! Her routine went as follows: filter in, cover down, test for five seconds, red light, remove defective filter and holds in right hand, insert next filter, cover down, press foot pedal, applies personal workmark to last filter tested (i.e. defective) and places in accepted tray, green light on, cover up, removes filter (which is to specification) holds in right hand, inserts next filter with left hand, cover down, press foot pedal, places the good filter into the reject tray!

I stood unbelieving and waited for a few minutes for another reject and the inspector did the same again. Failing concentration had caused the inspector to go out of step! End of true tale.

In fairness it was impossible to maintain such concentration. We modified the machine so that a red light locked the cover and caused another action to be taken before the cover could be taken up, to overcome the problem of day dreaming.

Just think next time you see some poor soul in purgatory standing by a conveyor supposed to be picking out defective items (be it biscuits, bottles, floor tiles, chocolates, pills, potatoes), what percentage are they really going to find? Are they likely to find more at 8.00 a.m. or 11.00 a.m? Are they likely to find more or less feeling fragile on a Monday morning, following a hectic Sunday night ?

What is the alternative to 100% inspection?

The circle of quality and the principle of quality assurance (the quality loop)
Figure 3.6 shows how quality is built in to the product or service at each and every step. This shows the common sense approach with 'Quality', 'Customers requirements' and 'Management Responsibility' interfacing and influencing each step of the total process.

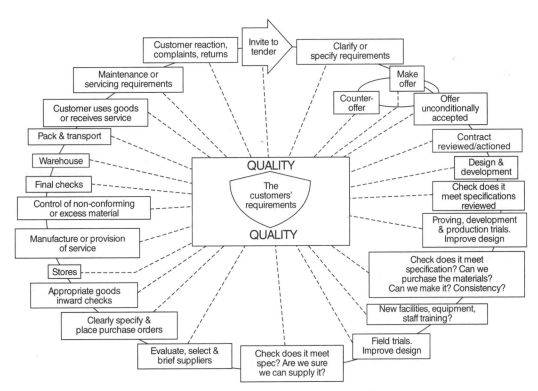

Fig. 3.6 Quality built in at every stage by reference to the customers' requirements

It also shows, the customer's requirement as a 'shield' at the centre of quality. I know this is stating the obvious, but possibly then it will stick firmly in your mind that this is the basic truth upon which all quality assurance principles should be built. This will also serve to reinforce the ISO 9000 definition of quality which we cover in the next section.

In the larger company the quality input will be made by a dedicated Quality Assurance Department.

In the very small company all this input must be made by the managing director or chief executive. The purpose of the model Quality Manual and Operating Procedures (see Appendices D and E) is to allow the MD or chief executive of a small firm to provide this input as appropriate at each stage, simply and routinely.

Definition of Quality

This is an area that can cause much confusion and you should have this definition very clear in your mind throughout and after your ISO 9000 project.

The confusion arises because there are at least four concepts or ideas about quality in a very general sense in people's minds or implied in the English language:

- The Rolls-Royce ideal;
- Fitness for purpose;
- The irrational concept of quality;
- The conformance concept of quality.

The Rolls-Royce concept of quality

You may consider that a Rolls-Royce is a better 'quality' car than a Skoda. This is not necessarily true. What is likely that many people may consider that a Rolls-Royce is a better 'grade' or 'style' of car than a Skoda.

You will hear people who have this concept of 'quality' in mind say, 'ISO 9000 does not guarantee a better quality of product or service'.

Fitness for purpose

Also expressed as 'meeting customer expectations' or 'ensuring customer satisfaction'. This may be useful if you actually know and have agreed what the intended purpose is. For some years I was the QA manager of a large valve firm which produced over 14,000 pipeline valves a week in sizes from 8 mm to 500 mm diameter. With numerous different materials and linings, it was impossible to know each valve's intended purpose. Part of the 'joy' of the customer returns meeting and investigation was to discover and be amazed at just what purposes the customers had actually used some of our products for.

If you can read into each and every one of your customers' minds and know what s/he 'expects' and what will keep them 'satisfied' you could make a career of being a mind reader.

It is also true that a person's expectations can suddenly and dramatically change. I will give a silly, but true, example that may stick in your mind. I was the youngest in the family. When we were very young it was safe to go for a walk and a picnic with sandwiches and lemonade. Being the youngest, during the picnic I was always the last to have a drink or 'swig' direct from the bottle of lemonade. It invariably had breadcrumbs floating in the bottle. This did not concern me in the least as an innocent infant. When I was about 8 or so, I suddenly realised where the breadcrumbs had come from and that lemonade was available without various foreign bodies. My expectations suddenly changed and I've never had lemonade with breadcrumbs again.

Similarly and far more seriously if your competitor offers a significantly better product or service, yours will suddenly become totally unacceptable at the price you are asking!

It should also be realised that you can only guess what a 'company expects' or give satisfaction if you know the mind of the most influential person in that company. As we know people come and people go, and the 'company's' expectation may change as suddenly.

The irrational concept of 'quality'

People can be persuaded (or persuade themselves) that 'X' brand or company is better 'quality' than any other, e.g. the housewife insists on buying cooked chickens only from Marks and Spencer; the businessman who only buys Rover cars; the office will only buy 3M discs; the local handyman will only use Dulux paints, etc., etc.

I can remember clearly in my younger days when I was a pump engineer, or 'rotating equipment specialist', that in this large chemical plant we had standardised after much logical thought, price comparisons and evaluation of reliability and service history, on two UK manufacturers of pumps over a specified range of models. This gave considerable savings on spares holdings and allowed substitution of pumps in case of failures. We were putting in a new large plant under the direction of a hard nosed project director for the USA. We had to fight tooth and nail because irrationally he insisted on using someone else's pumps because their pumps were the 'best thing since sliced bread'. This USA make of pump may have been excellent but at that time they were virtually unknown in the British market, so there were no spares available. (Incidentally the manufacturer never did penetrate the UK market.)

The conformance concept of quality

The fourth concept is the one that I suggest you keep clearly in your mind. That is:

> 'good quality is conforming exactly to the agreed documented specification and/or drawings.'

This is the unwritten 'conformance' definition of quality that is clearly in every Lead Assessor's mind. Do not confuse 'quality' with 'grade' or 'style'.

Grade or Style:

- high price or value
- high status
- provides esteem
- often impressive visual appearance
- selected features
- not necessarily high quality, reliability or performance.

Quality:

- correct to agreed specification (conformance)
- not necessarily high value or grade
- not necessarily high status.

With the above in mind, you may wish to reflect on several highly desirable objects that you dream of owning. Are they really of good quality or just of very high grade?

What is obvious is that very few companies have the luxury of providing and selling their products or services on the basis of grade alone. The majority of firms must get the quality right.

It is also clear that you cannot manage your business by trying to guess what each individual, or corporate customer, expects or will keep them satisfied at any particular time.

Obviously you cannot manage your business to meet an irrational concept of your product quality, it would be like trying to manage a changing dream.

There is simply no alternative to managing the quality of your business, whether it be manufacturing or a service industry. **You must document the agreed specification and then ensure you meet it.** Sometimes this specification will be clearly given by the customer in their contract documentation; this is the ideal situation that ISO 9001 visualises.

However, if your customer does not clearly specify what they want or their requirements are more generally implied (e.g. the standard of food in restaurant or cafe, the purchase of 'off-the-shelf' proprietary goods), you must clearly set out in simple written language (e.g. document) your own internal specification and then meet it. If you don't do this your staff will be unable to give a consistent service or product; they are not mind readers.

When you change the quality of your goods and service it is essential to change your agreed documented specification. It should therefore be clear that quality is not a relative item i.e. either high quality or low quality.

It either conforms to specification and it is therefore a quality product or service, or it does not conform and is defective. By virtue of this simple conformance definition, quality becomes **measurable** and hence **manageable**.

Remember in order to meet ISO 9000 and also to effectively manage your business the definition of quality is: 'good quality is conforming exactly to the agreed documented specification and/or drawings'.

Note The 'fitness for purpose' idea or concept can be a very useful loose measure when on the early feasibility or outline design stages of a project, to allow the target design specification to be drafted. However, this definition can be a disaster when the product is under manufacture (or service actually being provided). Every time the product is rejected the production manager or service provider will claim it should not be rejected because it is 'fit for purpose' or 'doesn't affect its functioning' etc., etc. When such items are then returned it will be solely the QA manager or inspector's fault for not rejecting them. Also your staff will become totally confused about what is acceptable. What is rejected this week becomes acceptable next week under different pressures.

A 'fitness for use' definition adopted when the product has started manufacture (or providing the service) is a recipe for customer complaints, customer rejects and loss of future orders. You have been warned!

Chapter 4

The programme or steps leading to successful registration

In this chapter I have detailed the programme, broken down into simple steps, that will lead to successful registration to ISO 9001, but excluding design. This is the most common requirement. Those organisations whose activities include a design element and therefore requiring ISO 9001 including design are advised to continue with the notes and preparation as shown and then simply add the requirements of 'design' (one clause of the standard only) using the notes within the additional guidelines in Appendix F.

These steps are shown in a realistic time-scale in Fig. 4.1. In my experience it is very difficult for this project to be completed and successful registration achieved in under six months. On the other hand even with very large sites or organisations with several sites it should **not** take over 15 months. If it is going beyond this time scale you would be well advised to review the situation:

- Look at the content coming out: is it over-documented? Have you a consultant who is making his fee larger by inventing and selling days?
- Have you an internal consultant or QA manager who is building an empire and writing himself into a job (often auditing the company to death)?
- Are you sure the person or consultant leading the project, really knows what he is doing?
- Is the leader of the project or other staff receiving any support or back-up; do they lack motivation?

Step I: Understand ISO 9000 and the Quality Policy Manual

If you refer briefly to the first figure in Chapter 5 (Notes on Writing the Quality Policy Manual) you can see where the manual stands in the hierarchy of documents. Appendix D contains a complete draft model of a typical Quality Policy Manual for you to use as a template to adopt and modify and make your own.

This Quality Policy Manual, when complete, addresses each clause (or section) of ISO 9001 in relation to your business. It is, in effect, ISO 9001 interpreted and written for your own particular organisation and how you intend to manage your business.

The Quality Policy Manual is a statement of your company's policy. It documents and makes public your commitment to the various requirements with respect to each clause of ISO 9001.

This policy document is the firm foundation onto which you build your operating procedures and the whole of your quality system.

	Jan	Feb	Mar	Apr	May	Jun	Jul	Aug	Sep	Oct
1. Basic understanding ISO 9000	▮▪▪▪▮									
2. Understand your business in relation to ISO 9000	▮▪▪▪▮									
3. Draft the quality policy manual	▮▪▪▪▪▪▪▮									
3. Inform/explain to senior staff	▮▪▮									
3. Inform & explain to all staff		▮▪▮								
4. Identify the minimum number of procedures		▮▪▮								
5. Draft the first procedures		▮▪▪▪▪▪▪▪▪▪▮								
6. Implement the first procedures				▮▪▪▪▪▪▪▮						
5. Draft second set of procedures				▮▪▪▪▪▪▪▮						
6. Implement second set					▮▪▪▪▪▪▪▮					
5. Draft third set of procedures					▮▪▪▪▪▪▪▮					
6. Implement third set						▮▪▪▪▪▪▪▮				
5. Draft fourth set of procedures						▮▪▪▪▪▪▪▮				
6. Implement fourth set							▮▪▪▪▪▪▪▮			
7. Document & implement fully								▮▪▪▪▪▪▪▪▪▪▪▪▪▪▮		
8. Standing work instructions (if required)						▮▪▮ ▮▪▮ ▮▪▮				
9. Consolidate, internal audits						▮▪▪▪▮ ▮▪▪▪▮ ▮▪▪▪▮ ▮▪▪▪▮				
10. Choose your assessment body				▮▪▪▪▪▪▪▮						
10. Pre-assessment visit							▮▪▮			
11. Plan the assessment								▮▪▪▪▪▪▪▪▪▪▪▪▮		
12. The frightener talk								▮▪▮		
13. Final preparation									▮▪▪▪▮	
14. Successful assessment & registration/certification										▮▪▮
	Jan	Feb	Mar	Apr	May	Jun	Jul	Aug	Sep	Oct

Fig. 4.1 Steps to successful assessment/registration

Step 2: Understand your own business, in relation to ISO 9001

When starting this project it is very common to hear chief executives or managing directors of small firms say that they have no systems or procedures. In all the projects I have carried out or managed I have never found this to be the case in an established company. No matter how small, they all have standard routines that have been based on previous experience of the staff, or by a simple process of trial and error, become custom and practice.

Let us briefly go through the clauses of the standard in simple everyday language and see how many of your informal procedures or routines already exist to satisfy these. They may not be written down (documented and formally issued) but I would be surprised if you haven't 60–70% already in place. ISO 9001 is no more than applied common sense.

Note Before you say, 'I have no procedure for that', or start to document what you think happens, or worse still what you'd like to happen, do check with your personal assistant, office manager, workshop foreman, etc. because it is common to find that in the informal situation the chief executive is not completely aware of the small informal subsystems which people have invariably introduced to make their own activities consistent and their life easier.

> A lot of people pick up silly non-compliances on audits because they haven't had the common sense or politness to ask the opinion of the person who does the actual task. REMEMBER, if Mary has been doing the job for the last 10 years, Mary is the Expert!

If you are managing and running an efficient business you are already half way there!

How many of the following aspects of your business are already in control, in relation to the basic requirements of ISO 9001 to manufacture your product or deliver your service? (e.g. Clause 7, 'Product Realization' and Clause 8 'Measurement analysis and Improvement').

Planning. (Clause 7.1, 'Planning of Product realization')
- You systematically plan and document your preparations for new contracts or processes and do not muddle through.

Contract review (Clause 7.2, 'Customer-related processes')
Before you make a tender, offer a service or accept a job, you ensure that:

- You know what the customer wants or expects and have communicated as required and have all the necessary information.
- You can provide the goods or service or have suppliers/subcontractors you know who can assist as required.
- You have a system to control and record 'Who' makes an offer and also 'who' accepted a contract.
- If it is a confirmed contract, in response to your tender or offer, you ensure that it is identical to what you actually quoted for.
- You want the job or wish to provide the service.
- If the contract is amended or changed, it is reviewed to ensure that you accept the change. If it is accepted you ensure that every member of staff who needs to know is told.

Purchasing (Clause 7.4)
- For purchased materials, services or items where it affects the quality of final goods or service, you use only suppliers you have evaluated and/or who consistently supply good quality material or services.
- You can demonstrate why they are approved by yourself and the approval is based on sensible reasons.
- That each purchase order adequately and uniquely identifies what service you need for the goods/material/service supplied.
- There is enough detail to ensure there is no ambiguity.
- That each purchase order is checked and authorised before being placed.
- That each delivery is adequately checked before being placed in storage and/or authorised for use.
- That, where appropriate, you have made arrangements in your purchase orders to allow yourself or your customer to inspect goods at the suppliers and/or to inspect the raw material or components at source.

Manufacturing, assemble or providing the service (Production and service provision. Clause 7.5)

- When you manufacture your product, or provide your service, you have effective controls that ensure your goods/service always meet the contract requirements and/or internal specifications and hence customers' expectations.
- That, where required, the controls are written down in simple language and are signed, dated and controlled.
- Key production equipment is given suitable maintenance.
- As required, you inspect and test the goods whilst they are being manufactured or the service as it is being provided;
- Within your organisation you control and can clearly demonstrate what is doubtful material (i.e. on 'hold' or quarantined), material awaiting test, reject material and that which is good.
- as required, there is a final inspection to ensure the goods/service complies with the contract and any records or paperwork required is provided.
- You have appropriate controls for the handling, packing, storage, protection, preservation and delivery preventing damage or deterioration.
- Where it is necessary, you will have a simple control system such as labelling or job tickets, so you can positively identify material or components.
- If it is a contract requirement, you will have a system that provides traceability to the original suppliers' material certificates.
  (Note: if traceability is not a specific contract requirement, ONLY provide if you think it is essential or highly desirable. Traceability can be very expensive!)
- Similarly, if necessary, you will produce items in batches or lots with a traceability system and records.
- If a customer supplies you with materials, services, staff, equipment, information etc., to assist you to fulfil the contract, you control these items to ensure they are not misused, lost or deteriorating.
- Where you have designed or manufactured a product and the client wants you to service it (or provide servicing instructions) you have effective systems to ensure these are provided to specified requirements.

Control of inspection, measuring and test equipment (Clause 7.6)

Appropriate test or measuring equipment is checked and controlled to ensure and demonstrate that it can test or measure to the appropriate tolerances.

Measurement, analysis and improvement (Clause 8)

- You measure and control your process, management systems and the end result to ensure conformity to requirements and to seek improvements. If the product or service needs to use techniques that involve statistical theory or practices, these will be used and controlled.
- Material or components that do not meet the required specification are controlled to ensure they are not inadvertently used or supplied.
- Such defective material or components are reviewed before scrapping, reworking, regrading to 'seconds', raising concession, etc.
- When there are defective materials/services and/or customer returns or complaints, appropriate action is taken to provide an immediate solution to prevent re-occurrence.
- If practicable you take actions to prevent problems before they occur.
- Each part of your management systems is checked independently to ensure procedures are being followed and are effective and they ensure that products/services consistently conform.

I trust that, having read the preceding very simplified summary of the requirements of the standard, you are saying to yourself, 'Well of course I do most of that or I simply would not be able to supply any goods/services'. It is quite obvious that a firm which accepted any contract offered, purchased from poor suppliers, or who did not control their processes would not be in business for very long.

If you are managing and running an efficient business you are already half way there!

The other items from the standard that have not been mentioned are:

The requirements to have your management system formally documented. Also appropriate controls of quality records to demonstrate a working QA system and to provide legible records of tests and inspection, where appropriate. (Clause 4 'Quality Management System' e.g. general requirements, for system, documents and records)

and

The requirements of top management to have comittment, be focused, demonstrate planning and control and have policy for quality and continual improvement. To have formal management reviews and to ensure that adequate facilities and resources are available. Also, to ensure training requirements are recorded, reviewed and training is provided where necessary. (Clause 5 'Management Responsibility' & Clause 6 'Resource Management')

Step 3: Draft the quality policy manual
First, decide who is going to do it. Then please take the Appendix D and with a red pen modify the words to reflect your business. Also refer to the notes in Chapter 5.

This should not be approached with any fear or anxiety as this is only a first draft. It is not 'carved in stone', it is likely to be amended several times before you are happy with it and before it is finally ready for assessment.

Note It is a common mistake to think that once you are successfully assessed and registered, your Quality Policy Manual, Operating Procedures and any Standing Work Instructions are in some way fixed. This is absolutely wrong as these documents will change and improve as you strive to improve your business, your management controls, the quality of your product/ service and as you introduce new products. Indeed, as an assessor, if you visited a registered company and their procedures hadn't changed in several years, you should be suspicious that these documents might be just a facade.

Note Do not be hesitant when drafting the Quality Policy Manual if you are in a **service industry** i.e. not a manufacturer of a tangible product. This often causes great problems with various industry guidelines and learned texts and papers. Do not get confused or misled. I am well aware that the standard was originally written and developed for Quality Assurance of businesses in the manufacturing sector. However, all you need to do is to think what is the service, or services, you provide and address them in the Quality Policy Manual and procedures as 'products'. If you can achieve this simple piece of mental dexterity and apply it to your quality systems and philosophy you will have few problems.

Typical 'products' for service industries could be:

- The care, safety and well-being of elderly patients.
- Installation, repair and maintenance of photocopiers.
- Provision of appetising meals, in a convivial atmosphere with clean plates, cutlery and tables. Ensuring that essential associated services are safe, effective and spotlessly clean, e.g. toilets, cloakroom, car parks.
- Maintenance of grounds, gardens and parks to customer specifications.
- Provision of security services and guarding of premises to agreed specification.
- Design and presentation of training or educational courses.

Note For educational or training establishments, I strongly recommend that you do not think of the 'product' or service as a person. It will lead to all sorts of trouble with 'rejects', 'corrective actions', 'customer supplied materials', etc. The 'product' should be the course or training programme.

Who is suitable to do it?

It is critical that you consider carefully who is going to document the system and who is going to implement it.

Only you can decide as only you know your own abilities, the time you have available, the complexity of your business, the ability of your staff, etc.

I would suggest caution regarding the following resources:

- Employing a retired senior manager. Will he or she document irrelevant things that they subjectively think ought to be covered or will they be objective?
- Not a project for a 'know it all' regardless of age.
- If you use a youngster they have got to be capable of extracting the input from others senior in age and experience. However, if you have a graduate (or similar) provided they have a sensible approach and excellent interpersonal skills, they could have an excellent grasp of your business at the end of the project.
- Not a project to give to a surplus member of staff because they will be generally regarded as a misfit.

Whoever you choose to document and implement your system (or you may choose to do it yourself) the person must have, or quickly develop, excellent interpersonal skills. Some people decide to use an outside specialist QA consultant. The reasons being to:

(a) provide expertise;
(b) provide 'arms and legs';
(c) provide focus and drive to completion.

This book provides (a) so think carefully before you call in outside help. However, do be aware that this project is going to take 6–8 hours a week of your, or someone else's, time for 6–8 months. See notes to Chapter 6, 'Advice on Consultants'.

Inform all your staff

> Having decided who is going to do this exercise, it is essential to inform all your staff who has been appointed and explain why.

Only you can make the subjective judgement how to do this, how big (or little) an issue to make of it. If not explained you might get people fearful for their jobs, thinking it is a work study exercise (especially if using an external consultant), or you can get a reaction from people not wishing to advise how they do their jobs as they may have built up an aura of indispensability or status about their job knowledge. A simple one-sheet memo on the notice board might suffice or you may need a formal meeting or presentation.

I have given numerous such formal presentations. I have never had an adverse reaction from the 'shopfloor' as they invariably see the sense of obtaining ISO 9000 to obtain more orders and job security, etc. Also they often prefer to have clear written instructions. The adverse comments and lack of enthusiasm will come from your junior or middle managers, especially those who excel at 'fire fighting' and thrive on confusion and lack of order.

Step 4: Identify the minimum number of procedures

You now have a draft of your very own Quality Policy Manual. Go through your Quality Policy Manual, clause by clause, and identify where you need a written procedure to ensure the requirements of each clause can be met. You should finish up with a list similar to that shown at beginning of Appendix E.

The aim, especially prior to assessment, is to obtain the minimum number of procedures. This will make the procedures easier to implement and also limit the assessor's target area on the actual day of assessment.

Having said the above, play safe, if you are not sure whether you need a procedure or not, write down in the list that you need a procedure. If in fact you don't really need it, this should become obvious during the implementation of the system. Look out for opportunities to reduce documentation.

Step 5: Start to draft and document the procedures

You should now start to draft the procedures using the examples in Appendix E as a guide. It is essential that the first two or three are chosen with care.

> Ensure you choose for your early procedures the ones that correspond almost exactly to current working procedures and which will not be controversial in any way.

You should aim to draft these first two or three and implement them with the minimum of fuss. You need an early aura of success. You need the staff to feel confident in you and also in themselves. They need to get the feeling that ISO 9000 isn't that hard at all, it's about documenting what we do. If anything, they want to feel that it's better than before. At least now it is written down and you know clearly what to do and what not to do.

Do not make the mistake of sitting in the office or in the lounge at home and writing what you think happens, what you hope happens or even worse what you would like to happen. This is a sure recipe for disaster as you will finish up with a set of procedures which are a facade and are almost bound to lead to a non-recommendation at assessment.

Ensure that you consult the true experts on the 'shop floor', especially if someone has been doing a particular job for several years. It is amazing how protocol often prevents discussion of draft procedures with the people who have to use them. Indeed if you can achieve it, the ideal situation is where your staff consider the formal documented procedures as 'their' procedures, how they carry out and control their own jobs. If they have an existing record book, proforma or rubber stamp that does the job, adopt it in the formal system. Improve their forms, make them look impressive, ensure they have enough space to complete the details, ensure the documented procedures are in a language they understand or even the colloquial language used. If some record is generally known to everybody as the 'green book' then call it the 'green book' in the documented procedures.

If people have already introduced a simple control of ticking or signing a form, e.g. signing a goods inwards note or copy of purchase order, to show goods inspected on receipt, etc., introduce a neat rubber stamp for them to complete signature and date, to properly formalise and reinforce their controls.

> **Note** Discourage a tick in isolation as a record of an inspection activity.

As an assessor I cannot identify and interview 'Mr Tick'. Encourage **signatures** (or **identifiable initials**) and **dates**. However, a long check sheet is acceptable if ticked off by ONLY ONE PERSON and then dated and signed at the bottom.

Do not worry unduly about getting the procedures correct on the first attempt. It can be expected and is the norm to amend these working procedures two or three times before they are entirely satisfactory. One or two procedures might have to be amended many times before they fully satisfy the requirements of the standard, or meet your requirements to efficiently run the business, or can be clearly understood by the least able member of the team.

Do not write procedures to impress the assessor. A very common mistake is to call up national, trade or international standards unnecessarily in order to give a professional image of working to the highest standard. This can be a grave mistake. Only call up standards if they are necessary or are a contract requirement, because it should be borne in mind that:

- You would have to hold a copy of every standard you have called up.
- You would have to control them to show they were the latest (or correct) issue.
- You would have to control the internal issue to your staff.

- You have just given the assessor an additional benchmark to check you against. 'You claim you are working to these standards, show me'. You have widened the assessors' target area and have given them more scope to find a potential non-compliance.

> The other common mistake, whilst trying to impress the assessor or your colleagues, is to use over-elaborate language. If your staff only understand words of two syllables do not use words containing three!

Do not employ technical language beyond the capability of your least able member of staff or you risk potential non-compliances arising.

You may hear it claimed that to satisfy ISO 9000 procedures have to be drafted in a certain format and contain certain critical words, etc. **This is not true**. However do be careful about using noncommittal words such as 'could', 'may', etc. See notes in Chapter 5 on writing the Quality Policy Manual.

Step 6: Start to issue and implement the procedures

A very common mistake is to draft and complete **all** the procedures and issue as a complete set on day X and on the next day expect everyone to be working to the new quality assurance (or ISO 9000) system. These are usually issued with a 'fear of death' memo or instruction that you must work to them or else, as they are personally endorsed by the managing director or chief executive who is committed to them.

If you do this, and **impose** new management procedures on your staff do not expect your staff to have any interest or commitment in your new procedures at all. No doubt they will do their best but if they are wrong and/or cannot be worked to, do not expect them to advise you. Similarly, with this high-handed attitude, if they know a better way of doing their job don't expect them to inform you.

The operating procedures must be drafted and issued as a continuous process, working hand-in-hand with all your staff. Informally issue one, two, or at the very most three procedures at a time.

> I recommend you issue your Quality Policy Manual and your first (and remaining) procedures in ring binders (or lever arch files) with the pages inserted into plastic or Rexel sleeves. This ensures they do not get ripped or heavily soiled. It also allows for ease of changing when amending/updating. I also recommend you print single side only and slip two pages in the sleeve back to back. When you eventually come to updating and amending you can easily get into a muddle with double-sided pages.

If you wish to embellish them it is very easy to produce blank pages with 'Quality Policy Manual' in large font on the top as a master. Also with a bit of 'cut and paste' with your photocopier (or with your DTP package) to put your firm's logo on the top as well.

You can do exactly the same with Operating Procedures. However, I tend to put these titles on the bottom of the 'master' page and possibly change the page numbering from a 'footer' to a 'header'. This enables one to very clearly distinguish the Manual from the Procedures.

A further easy, cheap but very effective enhancement is to print your Quality Policy Manual in one colour and your Operating Procedures in a different colour paper stock. (If you print your Operating Procedures on dark blue paper you make them difficult to photocopy, thereby making it easy to identify an unauthorised copy.)

Implement them, discuss them with **all** the staff associated with that procedure. Involve your staff (these are their procedures, remember). Improve them, simplify them, use appropriate existing informal systems that your staff have developed. Improve 'their' forms and Desk Top Publish them; give them better record books that look impressive. My own personal preference for a record book is the large old fashioned Cash Register type of book. If lined out properly and started properly staff can be persuaded to keep a neat and professional looking record book. The books because of their size are difficult to lose.

Any forms which your staff need (there are bound to be some, try and keep them to the minimum) should be Desk Top Published (DTP) to make them look as professional as possible. Do not worry if you or your secretary cannot DTP as any office bureau will do them for a small charge; a competent teenager with a home PC will do this for even less. I recommend that they should also be printed on good quality paper. If forms are photocopied make sure the photocopy is pristine. If you make the mistake of supplying poor quality forms, they are likely to be completed carelessly.

Note A coarse but true saying is 'Give men forms looking like toilet paper and expect the appropriate content'. This was brought home to me very graphically when I was a Production Manager looking at turn-over figures of £97million, with profit figures that intuitively based on experience were wrong. The accounts department insisted they were produced on a computer and hence must be correct. I pondered over these for some days (and sleepless nights).

I knew the materials/costs were carefully controlled, as were the production figures/targets as workers' bonuses were paid accordingly. Hence my doubts focused on the dispatch area, which was a remote, rarely visited railway sidings outside the explosive factory. The raw records were unbelievable! Scrappy notebooks, odd bits of paper, all completed or scribbled with stubby blunt pencils. I now believe all Computers should have a label 'Rubbish In – Rubbish Out'!

I recommend informality at the start to make the introduction as simple and painless as possible and suggest that you pull the documented system into formal control some three months before assessment. However, you may feel (and it might be more appropriate for your particular organisation) it is better to start off and to continue to issue all documents in a controlled manner. If this is your preference or it seems right for your organisation and staff, simply draft the operating procedure on 'Control of Documents' as the first procedure and work strictly to it from day one.

So issue and implement your procedures two or three at a time. Implement them, improve them, re-issue them, etc. When you and your staff are reasonably happy with them, issue the next two or three. This programme will typically take between 3 to 8 months depending on the size or complexity of your business and your commitment to the project.

Note At this stage, do not document procedures for products or services that fall outside your intended scope of registration. You do not necessarily have to have all your products and services registered. You may choose to include only certain products and services and consciously exclude others from the assessment scope.(However note: if you design the product or service, 'Design' must be included in your documented management system and also included within scope of registration. An important amendment included in the 2000 version of ISO 9001.)

Step 7. Document and implement fully

You have now after approximately 3 to 6 months documented all your operating procedures. Read carefully, review and walk through on your actual 'shopfloor' and ensure that they are all implemented fully. If not, amend, improve, simplify until you **know beyond doubt** that they are all working smoothly. You must be absolutely sure:

- all the staff are doing exactly what is documented;
- you have documented exactly what all the staff are doing relevant to registration.

When I state 'all the staff' that includes the managing director, chief executive and any other director, senior managers, etc. They are not exempt. In truth if anybody is going to try and shortcut the system it is likely to be one of the above! You may be aware that there are times when you have to take short cuts. If this is the case, document your short cut in your system with the appropriate controls to ensure that you don't foul the system and the product/service offered. You will note I have built typical common short cuts with appropriate controls into the draft Operating Procedures.

Step 8: Standing work instructions

What are Standing Work Instructions and how do they differ from Operating Procedure when required?

I have shown as Fig. 5.1 a very basic two-tier system. This shows all the processes controlled by Operating Procedures.

In larger organisations it is often necessary to put another third tier of documents under Operating Procedures, usually called 'Work Instructions' or 'Standing Work Instructions' and also occasionally with separate 'Inspection Instruction' or 'Quality Control Instructions'.

In a large organisation these then become:

- **Operating Procedures:** How we run individual departments or functions across departments (e.g. purchasing, training, etc.).
- **Standing Work (or Inspection) Instructions:** How operators are going to carry out one particular job.

In the majority of small firms it is not necessary to have this additional tier of separate Work or Inspection Instructions, it being far simpler to absorb them within the Operating Procedures.

However, you may judge (and only you can judge) that some jobs are so critical and the details so specific that some jobs must have specific Standing Work Instructions to have and to demonstrate effective control.

The mistake people generally make is to have too many or redundant Standing Work Instructions.

Before embarking on separate Standing Work Instructions consider carefully:

- Can I adequately cover the necessary controls within the Operating Procedures?
- Is what I'm proposing to document so obvious that it is simply not necessary?
- Does it affect the quality of the product or service I am offering to the customer, e.g. not washing one's hands after going to the toilet is not very hygienic but it will not affect the quality of the product in a welding factory. However, if the factory is making meat pies, such personal care would be essential.
- Can I within the training records show that the people doing these tasks are adequately trained so Standing Work Instructions are simply not required? Gardeners would not, in practice, plant a rose bush with one hand and hold an instruction book in the other; you must have records to show that the gardeners are adequately trained. Also you do not have to provide detailed instructions to motor mechanics on how to repair cars. They have training records to show they are trained mechanics. They can refer to the manufacturers' manuals as **guidelines** if and when required. Similiarly a trained electrician does not need an instruction to wire a 13 amp plug. Or, a graduate mechanical engineer, an instruction on how to stress a simple beam.

From the above you can gather that in the majority of cases detailed Standing Work Instructions should not be required. The necessary controls are provided through a combination of the Operating Procedures with the Training Records.

If you feel certain jobs must have Standing Work Instructions they must be controlled as per the Document Control Procedures. For guidance, two examples of Standing Work Instructions have been included within the model Process Control procedures in Appendix E.

If you have only one or two Standing Work Instructions it may be simpler to show and control them as appendices to the Process Control Procedure and not open another tier of controlled documents.

Step 9: Consolidate your systems and begin effective internal audits

You must now bed your system in. Yourself and your staff must truly feel that the Documented Quality System (DQS) you have jointly prepared as a team exercise is the only method we are going to use to run our business.

The following simple message must be clearly understood:

There are only two options; work exactly to the documented procedures; or change them to what we want to do.

You now start your cycle of **internal audits**. See Chapter 7 on Training Internal Auditors.

The secret for success at formal ISO 9000 assessment is for your own internal audits to be completely unforgiving and totally pedantic. Make everyone aware that the purpose of internal auditsis to find any fault and weakness in the system and then address it. Also to find any areas where you can improve or simplify. Do your friends and colleagues a favour by finding the faults in their system so that they can be corrected. Do not leave the faults there or the system untried to be found wanting and to fail the formal assessment.

Step 10: Choose your assessment body

You must consider this and choose your assessment body at least three months before you plan to have the formal assessment. This time is needed for you to submit questionnaires, the certification body to quote, and you to accept; then they will allocate an appropriate assessor who is likely to have a 2 to 3 month existing work load.

Do not make the common mistake of believing because the standard is published in the U.K. by the British Standards Institution that they are the only official body to be able to properly assess you and provide you with an accredited certificate. What is absolutely essential is that you are assessed by, and have a certificate from, a recognised and fully **Accredited Certification Body**.

UKAS publish a list of Accredited Certification Bodies (see Appendix A). At the present time there are approximately 60 bodies in the UK. Some are very small and lack status (even though they are accredited). Some cater for specialist sectors of industry e.g. ready-mixed concrete, steel construction, electric cables, timber, reinforcing steels, However, the majority of firms in UK will be registered with one of the following bodies (shown in alphabetic order), all with a wide scope of registration and expertise:

British Standards Institution (BSI/QA, not to be confused with BSI/Standards)
Bureau Veritas Quality International (BVQI)
Det Norske Veritas Quality Assurance (DNVQA)
Lloyds Register Quality Assurance (LRQA)
National Quality Assurance (NQA)
S.G.S. Yarsley International Certification Services (SGS).
United Registrar of Systems (URS)

There are similar recognised bodies of status in other countries, although all of the above offer and provide certification throughout the world.

My advice in previous versions of the book, was that there was very strong competition between the above bodies and by going to three of them for a quote you will get a fair price. Unfortunately this is no longer the case. Some have gone for a policy of 'the product is too cheap' or 'We'd rather have 3000 registered firms at £900/day, rather than 5000 registered firms at £600/day'. Hence do expect a wide variation in prices quoted.

Also some have very large prestigous office blocks in prime locations, fancy expensive brochures, large marketing or training divisions, etc. This will reflect in the fee, but consider does it give you any added value?

Do not under any circumstances accept a certificate from an unaccredited body as it may be judged to be unacceptable.

When going to the above accredited certification bodies, write or telephone/fax them and ask for one of their application forms. Do not bother writing a long letter describing what you do, etc., as they all have designed simple forms to ask the questions they need to know.

> **When you return the form, please make it clear on the form or the covering letter that you require a fully accredited certificate.**

This is essential. These certification bodies do not have an across-the-board accreditation. They only have accreditation for areas where they can demonstrate adequate experience. At least one of the above has issued many certificates without telling the client that they are not accredited for that particular activity and have issued certificates without the DTI 'Tick and Crown'. In a few cases the certification body may need two or three assessments in your type of business witnessed by the UKAS to obtain accreditation. This may be acceptable to you but they should tell you as part of the offer quote.

Things for you to bear in mind when comparing quotes from certification bodies:

- Does one body give me more status with my customers or in my market sector? Is this real or imaginary? Is this certification body's logo acceptable and an advantage in all my overseas markets e.g. 'British', may have imperial or colonial overtones? Is an UKAS accredited ISO 9000 certificate from any of the above equally acceptable?
- What are the objective criteria that their assessors will be assessing me against? What is their pass rate at first attempt at assessment?
- What is the true total cost? Please work it out very carefully – some have additional certification costs, some have an extra triennial assessment with very significant additions (which they may hide by quoting over a three-year period), some charge for travel and expenses. URS include certification, travel and expenses and all costs in the assessment and annual fee.

Use a table similar to Fig. 4.2 if you are going for a competitive quote. You must ensure you are comparing like with like.

- How many person days for the assessment and which or how many sites will they visit, if you are a multi-site operation?
- Do you feel comfortable with a long-term partnership with the certification body. Do they seem superior, remote, bureaucratic or do they seem friendly, approachable and treat you as a valued client?
- Are the people you are now talking to the people with whom you will be dealing in the future or just a sales office?
- If you feel you need it, ask if they will visit you. Do not think the one who automatically offers to, or insists on calling to see you is the best. They may be the worst, indeed if they provide free visits to every potential client they are likely to be the most expensive!
- Will the certification body work to your assessment timetable and not impose their own?
- Are the surveillance visits announced (no surprise visits at inconvenient times to clash with other visitors, work schedules, holidays, etc.). Is there some flexibility that the certification body will bring annual surveillance visits forward or put back a couple of months if your workload, projects, etc. demand it?

Complete your cheque and send the application form off to your chosen body.

Step 11: Plan and stage manage the assessment
Firstly go through your Quality Manual and Operating Procedures, Work Instructions, forms, schedules, etc. and make absolutely sure all the cross references and issues/dates are correct. It is highly likely that, having developed this system for 6–8 months, some items are now not consistent.

Name of Certification Body			
Application fee			
Fee for full day pre-audit visit & on-site document review			
Fee for assessment			
Fee to prepare & release certificate			
1st year annual certification fees			
Cost of 1st year's surveillance			
Estimate of additional cost of travel & subsistence			
2nd year annual certification fees			
Cost of 2nd year's surveillance visits			
Estimate of additional cost of travel & subsistence			
3rd year annual certification fees			
Cost of 3rd year's surveillance visits			
Estimate of additional cost of travel & subsistence			
Cost of triennial assessment at end of 3rd year			
Estimate of additional cost for travel & expenses			
TOTAL (3-year cost)			
Average monthly cost (÷ 36)			

Fig. 4.2   Work out the true cost of your certification

If you have not done so already you must now enforce the 'Procedures for Control of Documents'. This must be done rigorously. It is surprising that one of the largest causes of non-compliances at assessment is minor errors in controlling documents.

At about this time, approximately 3 to 4 months before the actual on-site assessment, write a memo to all your staff on the following lines (possibly posting it on the notice board):

> 'As you are aware we have been working hard together over the last few months to fully document our Operating Procedures. I now believe that these are complete. As from .... (day) ..... (month) .... (year) these are to be complied with. It must be clearly understood that if we are to be successful at the forthcoming assessment there are only two alternatives: *Work exactly to the prescribed procedures, or request they be changed. Each time anyone fails to work to the prescribed procedures there is the likelihood of receiving a non-compliance and failing the assessment. Please do your best to ensure that this does not arise.*'
>
> (Managing Director)
>
> (Date)

This memo has two purposes, firstly to advise your staff that the exercise is now for real. Secondly, to put in an artificial cut-off date prior to which you will not accept non-compliances. If the assessor tries to raise non-compliances against practices or records prior to the date of the memo, refuse to accept them.

> This restricts the assessor's target area to records only going back 3 to 4 months prior to assessment.

It is also essential at this time you **change your whole attitude** and approach. Up until now you have been concentrating on getting effective, efficient and economical Operating Procedures. For this last run-up to the assessment, especially the last 4–6 weeks, your approach should change to one of 'we are going to pass this assessment, this is my top priority'. You must now motivate people to work exactly to the procedures and remove any potential causes of non-compliances.

You may at this stage wish to consider if there are any additional procedures that would be useful to run your business or give extra management control that you know is actually outside the scope of ISO 9001. Typically these may be accounting procedures, job-costing procedures, personnel procedures, Health & Safety, COSHH, HACCP controls, etc.

As I have just said, you know that, strictly speaking, these are outside the actual scope of 9001:2000 registration.

> However, you may judge (and only you can judge) that there may be a 'one-shot' opportunity to get these additional management controls or operational procedures documented and implemented.

Step 12: The frightener talk

If you do put in these additional procedures, modify them slightly and put somewhere in the introductory paragraphs 'this procedure is outside the scope of our ISO 9001 registration and is excluded from assessment'. That is, you have got it in and working but you have clearly taken it out of the external assessor's target area.

About three to five weeks before the assessment you need to give a preparatory talk. This talk should be a 'warning' talk so that if there is somebody not pulling their weight or not working to the procedures, or their procedures are flawed, they suddenly realise this is for real and they had better sort themselves out and get organised. Hence, the three to five weeks prior to assessment; three weeks is the minimum time that somebody will need to sort themselves out and get some effective procedures in place. Longer than five weeks away from the assessment this talk starts to lose its impact.

Such a talk should be given to **all** your staff. You may have to modify it for presentation to managers as opposed to 'shopfloor' staff. However, it is essential you brief 'shopfloor' staff. Don't blame an operative if they cause a problem on the day if they don't even know who the assessor is and are either terribly rude or over-helpful. (Believe me, it is truly awful when an operative starts chatting away and exposes all your weaknesses!)

The talk should cover the following items, (it also gives you the key items to consider to plan and stage the assessment).

OUTLINE DRAFT FOR THE PREPARATION FOR ASSESSMENT

. . . . . plc is going for assessment to be approved as a Registered Firm to the International Standard ISO 9001:2000.

Why is it essential for . . . . . plc?

- Mandatory requirement of customers
- Reduction of multiple assessments
- To survive
- It's good business practice
- To cut the cost of poor quality
- Protection against criminal prosecution and civil claims
- To give a marketing advantage

(*See notes in Chapter 2.*)

What is an assessment?

It will be a formal inspection by **trained** observers to certify that we have effective management systems that ensure:

- we always meet our contract requirements and our own specifications exactly;
- work is being carried out exactly to our own approved Operating Procedures;
- our management systems and procedures satisfy the requirements of ISO 9001 (within our declared scope).

You should note very carefully that this is **not** the normal customer quick courtesy call.

It should also be noted that this is not like a normal exam, where you build up marks and eventually pass. The opposite happens; the assessor starts finding 'non-compliances' (explain) and eventually you/we fail. The assessor does not award points for things done correctly; they can only raise non-compliances. Hence in practice they are basically coming to test and try our systems. If they cannot fault our systems then we will be recommended. Hence in practice you pass this assessment by minimising the number of non-compliances.

Up until now we have concentrated on getting good systems and procedures. For the next . . . . . weeks we must now concentrate on passing this assessment, **whatever it takes.**

When is it?

The assessment will take place on . . . . (day) . . . . (month) . . . . (year). Starting at 9.00 am.

Initially for about 20 minutes there will be a formal opening meeting with managing director and appropriate personnel.

The assessor(s) may then have a quick tour of the office/factory and familiarise themselves with our product.

Thereafter from 9.30 am you are subject to assessment by the assessor(s). The assessor(s) will **at all times** be accompanied by a guide. The assessor(s) may make visits to every department and throughout the whole site. The assessor(s) may ask questions of any member of staff, any employee or sub-contractor.

The assessment is likely to finish at approximately . . . . pm on the . . . . day. The assessor(s) will then take about an hour to write the report.

There will be a meeting with the managing director and appropriate personnel to present the findings of the assessment.

The appointed guides throughout the assessment will be . . . . . . , it is essential that they do not leave the assessor(s) on their own.

(In larger organisations the departmental head or nominated departmental representative should also be named and be on call and stand-by for the whole of the assessment period.)

How can we fail?

To fail us, the assessor(s) must demonstrate that we don't comply. To do this they must find **objective evidence** to show we don't comply, which will be raised as a non-compliance. As they do not know . . . . plc, any non-compliances, or failures of our system will be ones that we have allowed them to see and which we have exposed to them.

These non-compliances are raised as minor or major. We can get a few minors and still pass. One major non-compliance and we fail and have to go through this all again with, I should add, additional expenses and fees.

How can we pass first time?

Work exactly to contract and agreed specification.
Work exactly to the approved Operation Procedures (and instructions?).

. . . . plc must look ship-shape and professional at the start and all through the assessment. You must remove all non-conforming objects or attach suitable labels. You must remove all out-of-date drawings, forms, specifications, contracts, odd letters, etc. When I say remove, I mean remove. File them in their proper place. If they are irrelevant, seek confirmation and then destroy them. Do not hide them.

What is the assessor looking for?

It should be clearly understood that the assessor(s) attempt to find non-compliances:

- by examining and cross-checking documents in detail;
- by inspecting product, equipment and looking all around the site;
- the assessor will be observing and also asking questions. The questions may be asked against sections of the standard. The questions may be checking sections of the Quality Manual and individual operating procedures or work instructions. It should be clearly understood that the assessor can check any individual in any department;
- the assessor could also carry out horizontal audits, checking how one department interfaces with other departments;
- one of the most critical audits the assessor can carry out is a vertical audit, that is pick up an order that has just been completed, or ready for despatch, and then trace it back through the system, looking for inspection records, process control, purchasing controls, etc., right back to the contract review of the original order or tender. If the system is not fully implemented a vertical audit will quickly demonstrate that fact;
- the assessor will also follow leads that you provide, so don't get other departments or your colleagues into trouble by making unwise comments or stating unproven facts or opinions.

Do not think you will get away with a shallow answer to a trained assessor. They will always ask 'show me' or 'what if' so if you claim you have controls or records they had better be available!

How to handle an assessor

- Do not volunteer information. Do remember it is their job to find non-compliances not yours to point them out.
- If you are asked a question, think carefully before you answer. Ask for the question to be repeated if necessary. If you are still not sure what they are after, ask the assessor(s) if they could explain what they mean.
- If they appear dissatisfied with your first answer do **not** change it or volunteer more information. Your first answer especially if you have thought about it is probably right.
- Do **not** anticipate the questions. If they do not ask for something or do not ask a question, do not anticipate it. I repeat **do not**.
- Similarly if the assessor is looking at your records or standing at your workplace, and is silent, do not start explaining anything: wait until they ask. If you can't stand the strain of a long silence and you feel you must break it, start talking about football, knitting, television, etc.
- Do not argue with the assessor. Wherever possible leave the guide to answer queries, etc.
- Do not start to argue or disagree with our guide in front of the assessor.
- Do not lose your temper, criticise the firm, criticise your colleagues, start on about your 'hobby horse', etc.
- During the assessment under no circumstances (other than Health and Safety issues) should you distract colleagues or the guides from their assessment task or advise them of current problems in front of the assessor. You must wait until the assessor has left the area.
- The way to succeed and to obtain ISO 9001 on the first attempt is to remember, that the over-riding priority during these days is passing the assessment.
- Basically the correct answer to give the assessor is on the lines of 'I'm following the approved written procedures and instructions'.

- The Operating Procedures and instructions must appear as being used day to day and an integral part of normal activities. They must appear to be used all the time, i.e., immediately available, obviously 'thumbed' not brand shining new.
- You must read and know the Operating Procedures for your job 'inside out' and 'back to front'.
- There are only two alternatives to succeed in obtaining ISO 9000 registration: work strictly to approved procedures and instructions; or get them changed to reflect the preferred system.
- Please try and make the effort to be polite and friendly to the assessor(s).

Step 13: Preparing for an assessment

Do not simply drift into the assessment. Below is a checklist you may wish to use (or give your guides) to check approximately a week before the assessment. Also, most important of all, the day before and each morning of an assessment, come in early and re-check.

Checklist for guides and departmental heads

1. Have all obsolete or uncontrolled documents been removed?
2. Has all unidentified material been removed or labelled?
3. Has all uncalibrated or unauthorised equipment been removed off the site?
4. Is **all** filing and records up to date?
5. Ensure that **all** drawings, specs., procedures, work instructions, are clear, readable, correct issue and where appropriate they are also authorised and dated. Preferably there should be no manuscript ammendments. If there are, ensure they are also correctly authorised(e.g. signed and dated).
6. Is the identity and inspection status of all work clearly visible?
7. Are the authorised Operating Procedures, work instructions or other appropriate information readily available to the staff who need it?
8. Do the assessment guides know the organisation's Quality Manual and Operating Procedures? The guides should know them in absolute detail.
9. Does **everybody** know their **own** procedures? Are they following them **exactly**?
10. Do you have prepared sets of contracts, drawings, processes, etc. that you have completely checked, just in case the assessor gives you the choice to present a sample of your choice.

Additional advice for the company guides

- You may voice disagreement or offer further explanation to the assessor(s) but do not argue with or hassle them.
- If the assessor asks a question of an employee which you know is not within the employee's job description/knowledge you must cut in and interrupt immediately and should tactfully advise the assessor of this fact.
- If the assessor decides to complete a non-compliance form he will ask you to sign the form to verify the factual statement made, which must be supported by objective evidence.
- Do not hurry an assessor. It is his/her prerogative to vary his/her programme and also it is his/her responsibility to keep to their planned timetable. If they run out of time and do not see all your departments it is to ..... plc's advantage.
- Consider staff suitability. The last thing for you to consider carefully is if all your staff are suitable for assessment.

There are certain types of people that you would be advised to keep clear of an assessor:

- the person who loves an argument;
- the 'know-all' or 'clever-dick'.
- the person who simply cannot keep their mouth shut
- the person who moans all the time about the firm or cannot stop 'riding their pet hobby-horse';
- somebody of low communication ability;

- somebody with a low stress threshold;
- someone with an aggressive attitude, etc.

Do ensure that your official guides are not in any of the above categories or you will have a disastrous assessment.

Causes of non-compliance

A non-compliance should only be raised against:

- not working to your customer's requirements or own internal specifications;
- staff are not working to your own procedures (or instructions, where applicable);
- your Quality Policy Manual or Operating Procedures do not satisfy the **words** of ISO 9001:2000
- not meeting the certification body's regulations (this item included for technical correctness only, most unlikely to arise).

If the proposed non-compliance does not meet one of the above criteria or is not supported by objective evidence, ask politely for further explanation of what the non-compliance is actually against before you can sign to agree or confirm. (**Note** If it does not fit into one of the above catogories, 'stick by your guns' and quietly but firmly challenge the assessor!)

'Dummy' assessment by an outsider

You are now very nearly there. If you need that extra confidence, one option is to get an experienced assessor to carry out a semi-formal assessment.

If you are a small firm this should only take one or two days. The advantages it can give are:

- experience for yourself and your staff of being assessed and the assessment process;
- an independent outsider to check your work. It is true that you never see your own mistakes no matter how long and how often you check your own work;
- remove actual and potential non-conformities before the formal assessment;
- increase the probability of a first time pass;
- if done 'professionally' with adequate records and covers your whole QA system, it can be used as evidence for your first internal audit.

One sensible option taken by small firms is to sub-contract this 1st dummy (internal) audit to QA consultants and then place a maintenance contract on a QA consultancy to do their internal audits for them. If you do consider this second option, for straightforward firms of up to say 25 employees it should be possible for a trained and experienced auditor to completely audit your system in 2 days per year.

If you do decide to carry out the option of a dummy assessment you need a good consultant (see notes on Chapter 9). A similar effect can be obtained by requesting your certification body to carry out an extra day visit to carry out a 'dummy' assessment, usually called a 'pre-assessment' visit. This is usually a slightly more expensive option but should have the advantage of having the same assessor carry out the pre-assessment and the formal assessment. Obviously you should make this a condition of the 'pre-assessment' visit.

Note However do consider, the overall cost of having a pre-assessment visit followed by an audit. In total the cost will be about the same as failing and then having a re-assessment (and if you pass you have halved the cost).

Step 14: A successful assessment and registration

Well you are nearly there! Just that final bit of preparation and effort and you can proudly say that your organisation is among the best. Not just because you say so, you will have a certificate following a formal assessment from an officially approved body that says so.

Items for you now to consider on the actual assessment day(s) follow.

1. It is vital you stress to everybody that the most important thing on this day(s) is passing the assessment. It overrules everything (other than health & safety issues).
2. Do the assessor(s) know how to find you? Do you need to post or fax them a map?
3. Do they need help in finding accommodation? Do not offer to pay the hotel bill of the assessor(s) as it will be rejected and will cause offence.
   (unless this has been formally agreed and built into the contract).
4. Does the assessor need picking up at the airport or station?
5. If they are coming by car have you made arrangements with the gatehouse to let them in and have you reserved a parking space?
   Ensure that the receptionist knows their name; go and fetch them personally from reception. (It makes sense to make the assessors feel like important guests and get them on your side.)
6. Walk around last thing the night before. Get in early in the morning and walk around to ensure the housekeeping is pristine and that there is no silly non-compliance in evidence, e.g. unidentified material, uncontrolled non-conforming material, work in progress without proper authority or procedures/instructions, etc. Do be observant and look critically. It is surprising the silly things you may see.
7. Have tea/coffee and biscuits available for the opening meeting. Have the chief executive present at the opening meeting, to demonstrate commitment at the highest level, to welcome them to the company. Give an outline background to the company and say how much you are looking forward to the assessment. Introduce other members of staff present, which should include the guides. Then sit back and let the Lead Assessor chair the **opening meeting**. The Lead Assessor should:
   - confirm the assessment scope i.e. what products or services are included within the Quality System for which you are seeking certification;
   - introduce other assessment team members (if any);
   - outline the assessment programme, including the approximate time anticipated for the closing meeting;
   - explain how non-compliances and observations will be raised and recorded;
   - confirm that any information is strictly confidential;
   - ask if there are any Health & Safety items or Trade Union issues of which he should be aware;
   - confirm administration and lunch arrangements, request or confirm who is the guide and ask for photocopying facilities, etc., if required.
   - Any questions?

   Having completed his meeting, the assessor may or may not want a quick tour before starting assessing your system. If he does so then again this is your chance to impress the assessor. Ensure the managing director personally shows him around before handing over to the guide.
8. If at all possible give the assessors an attractive little office as a base with table and chair and a phone. Make them feel very welcome, important and comfortable. Feed them unlimited supplies of tea/coffee and biscuits!
9. On training courses for auditing they will cover assessor tactics and the tactics of the organisation being assessed. One of the auditee's tactics might be to waste the assessor's time. This is not recommended with any Lead Assessor from the UK accredited certification bodies. To be registered as a Lead Assessor in the UK you have to be very experienced. They are likely to spot this tactic a mile away and it could be very counter-productive.
10. Ensure the guide is suitably briefed (perhaps tell him to read these notes two or three times) and stays with the assessors all the time. This is essential, to stop the assessors wandering and asking questions at random. Also the guide can maintain some consistency of approach. If an assessor has no guide it can sometimes happen that the line of questioning or the answers being given will give rise to inconsistencies and hence non-compliances. The presence of a skilful guide can prevent this. Also a key job of the guide is to interrupt and cut anybody off who is asked a question that is outside their job knowledge and responsibility. This must be done immediately, even in

mid-question, to prevent somebody who does not know, and does not need to know, doing their best to answer or giving an inspired guess.

The selection of the person to be the guide can be vital. The guide **must** have the following qualities:

- They must be well known and respected throughout the organisation.
- They must be senior enough to command respect, but not necessarily a very senior executive or manager.
- Must have excellent interpersonal skills as, with the assessor, they may be talking to and need the full cooperation from any grade of person in the organisation;
- Have expert knowledge of your product and/or service;
- Know your Documented Quality System inside out. (It does not impress an assessor at all when the conversation goes 'I think there is a form in our system to cover that point – I can't find it – can you see it?' or 'whereabouts in our Manual/ Procedures do you think we will have (or the consultant has) covered that requirement?')
- Remain calm under stress, will maintain concentration, will not get irate or lose their temper, panic, etc.
- At appropriate times can keep their mouth shut. Will not open their mouth until their brain is firmly 'in gear'.

11. Confirm arrangements with the assessors for lunch. The assessor should normally fit in with the normal time of lunch break for your organisation. Do offer the normal courtesy extended to visitors of providing food at lunch time. It is likely that the assessors will turn down your offer to go out for lunch and request sandwiches or a quick bar snack or local cafe. If the assessors do agree to go out for a full lunch taking an hour or more, say nothing and entertain him. On this day, regardless of what you may think, the most important thing is to pass the assessment and get recommended for registration.

12. Remember the aim of this day is to pass the assessment.

You pass the assessment by **minimising non-compliances**.

What to do when faced with a non-compliance (or n/c)

The assessor should discuss what has been observed and why it is a non-compliance.

If at all possible you do not want this or any non-compliance. Remember you pass by minimising non-compliances. The guide should think this through before agreeing to sign, or better still start discussing earnestly before the assessor writes anything down.

What is a correct and genuine non-compliance? It should have two parts: (1) It must be an observed fact (or factual omission), not the assessor's opinion or supposition; (2) a statement of why this observed fact or omission is a non-compliance (n/c), which should be one of the following reasons:

- The observation shows that you are not providing goods/services to your clients' contract requirement or you are not complying with the requirements of your own internal specifications. (**Note**: if neither of these 'requirements' could be ascertained you can fairly expect a non-compliance.)
- The observed fact (or omission) clearly demonstrates you are not working to your own approved Documented Quality System (e.g. your own Operating Procedures/Instructions or process controls).
- The observed fact (or omission) clearly demonstrates you are not complying with the **words** of the requirements of ISO 9001.
- The observed fact (or omission) shows you are not complying with the published regulations of the certification body.

If the observation is a clear contravention of one of the above, sign gracefully! If your organisation has written in their procedures that they are going to do 'X' and they are not doing it, you are not in a position to object.

Sign the n/c, and if you've got the time and resources amend the procedure or controls in draft to discuss with the assessor and get his agreement that it clears the n/c.

If you think it is not a non-compliance (or may not be) do not start to ARGUE with the assessor, i.e. don't immediately construct an 'I win/you lose' situation. First of all decide if this is an observed fact (or omission) or just the assessor's subjective opinion/supposition. If it is the latter, decline to sign the n/c.

Ask the assessor to explain what this observed fact or omission is a non-compliance against. Ask him/her to educate you. You may comment: 'I don't think this is in our contract requirements, I don't think it's in our procedures. Is this requirement in the words of the standard?' etc.

If it is against your client's contract requirements or your internal specification, it is often possible for the guide to explain away the n/c by superior knowledge of the product/service.

If it is claimed as an n/c against the words of the standard, ask the assessor to show you where the requirement is in the 'words' of the standard. Sometimes the 'words' of the standard will work for you. I stress the n/c must be against the 'words' of the standard and not against the assessor's interpretation of the standard or his extrapolation of the standard.

If the guide can confirm it is a genuine n/c then the n/c should be signed.

If the guide does not agree it is an n/c then very politely decline to sign. This situation should rarely arise as the assessor should be under instruction that if there is any doubt at all they should not raise an n/c, although they may express concern by raising an observation.

Note Not all certification bodies allow assessors to raise observations. These are for the assessor to raise concerns or to advise the client of potential n/cs. They do not affect the recommendation in any way.

Despite their reputation, most experienced Lead Assessors do not find it essential just to raise at least one non-compliance just to show they have been there. However, it is unlikely that an organisation will go through an assessment with no non-compliances raised. It does happen but the percentage is small, probably in the order of 2% to 3%.

What is important is that any n/c raised is a minor non-compliance, and not a major non-compliance (or a 'hold point' in LRQA's jargon).

A **minor non-compliance** is an isolated lapse of the system e.g. one purchase order out of a sample of 20 not signed, one item of non-conforming material not identified or segregated, etc.

A **major non-compliance** is where there is a total breakdown of an essential element of the system, e.g. a whole procedure missing; one department or a key person in that department not working to the system; the company knowingly sending out material not conforming to specification. The formal way of actually defining a major non-compliance varies from certification body to certification body.

You can have several minor n/cs and still be recommended for registration. However, one major n/c and it automatically means the assessor cannot recommend you for registration and will mean at least a partial re-assessment (with obvious additional expense both internally to correct it and externally to pay the certification body's additional fee).

Whilst I have advised you that you can obtain several minor n/cs and still get recommended for registration, you should do everything prior to and during assessment to avoid minor n/cs.

If you obtain **several minor n/cs** against one clause of the standard, or occurring in one procedure or location/department, these can quite legitimately be used by the assessor as objective evidence to make **one major n/c**.

After the assessor(s) have completed the assessment they will adjourn to write and compile the report.

There should then be a **closing meeting** at which the assessors should:

- thank you for your courtesy and hospitality, etc;
- advise you of their recommendations;
- go through any n/cs and observations raised;
- restate that confidentiality is respected;
- state it has been a sampling exercise and there may be n/cs that may not have been found (e.g. he should 'state the disclaimer');
- if not recommending you, advise you of the right to appeal;
- ask you to sign the appropriate part of the reports;
- ask if there are any questions.

If you have documented your system correctly, if you are working to it, if you have minimised the potential for non-compliances, you can expect to be recommended for registration.

If the assessor states that he is recommending you for registration it is 99% certain you will be getting your formal certificate in two or three weeks.

If you have failed, think through why. If it is your fault, correct it quickly and apply for re-assessment as quickly as possible.

If, on reflection, you think the assessor has been unfair or the non-compliances are not genuine or just his/her subjective opinion, appeal in writing. It is a 'no lose' situation: either your appeal is successful and you are registered or it is not and you go for re-assessment as before. **Note** In the UK the appeals go to a completely independent panel or board e.g. outside the certifications bodies control. In practice, if you have a genuine case with objective evidence, it will be 'sorted' before it goes to the Board of Appeal.

Note   If you are not happy with the assessor, request another assessor for the re-assessment. I can assure you that this will make no difference at the re-assessment one way or the other. People are sometimes afraid to make this request. Remember you are the customer and you are paying.

Similarly, if you feel you would like a change of assessor after the re-assessment or a surveillance visit, do not be afraid to ask.

Having paid heed to all these warning notes the probability is that you've passed. How do you feel? I warn you now after the initial ecstasy, there will be a couple of days of anti-climax.

However, do be pleased with yourself, do tell your customers, do have a formal presentation of the certificate and **do remember to thank your staff**.

Step 15: Progress after registration

Take your system, continually improve it and, where possible, simplify it.

If you are a small firm, do not be rushed by some consultant, quality centre or government initiative into a Total Quality programme. Read Chapter 8 to get a vision of what Total Quality is about. Some Total Quality programmes are related to the 'culture' of large firms and of changing peoples' attitude.

The managing director of a small firm is 'the culture' and he sets the 'attitude' of the whole firm. Look at the realistic models of Total Quality shown in Figs. 8.1 and 8.2 of Chapter 8. Now you have your management or control, you can consider improving your people, process and product/service design.

If you are part of a larger firm read Chapter 8 and perhaps consider a Total Quality programme but a sensible one based on what can realistically be achieved.

Do consider the Environmental Standard ISO 14001 if you are emitting waste, scrap or by-products down the drain or into the atmosphere. There are real benefits to be gained and having obtained ISO 9001 you are half way there.

> In a small firm these Total Quality programmes are often a nonsense. A more practical alternative is to look at ISO 14001 Environmental Standard registration.
>
> As the requirements are aligned to ISO 9001 standard, it means you are already 2/3 of the way there.

The requirements of ISO 14001 neatly mesh in with your ISO 9001 system.

Note there is a sister book to this book *ISO 14001 Environmental Registration Step by Step*, published by the same company, written by Tony Edwards

One of the most sensible things a small firm can do is use your Documented Management System and add your personnel requirements and your Health and Safety responsibilities onto your Operating Procedures. Hence by using the same discipline you pull these under demonstrable and effective control. In the UK the appropriate standard is OHSAS 18001.

One acronym to hit the QA world is QUENSH management, which simply stands for QUality, ENvironmental, Safety and Health, which is based on sensible documented fully integrated management systems and practical controls as used for ISO 9001 approval. In more recent times this has been simply replaced by talk of having an 'Integrated Management System'. I now have a couple of clients who are registered to ISO 9001, 14001 and 18001 with only one Manual and one set of procedures. I'm sure the number will grow.

Chapter 5
Notes on writing the Quality Policy Manual

It must be clearly understood what the Quality Policy Manual is and what it is not, where it stands in the hierarchy of documents and its function.

The documented quality system

The ideal and recommended model for the hierarchy of the documented quality system (often referred to as the DQS) is as shown in Fig. 5.1. Although these are usually called the documented 'quality' system, in reality to be truly effective they should be the documented MANAGEMENT system, e.g. how the management of *all* departments ensures products and services are supplied to specification.

Quality Policy Statement

At the top is the **Quality Policy Statement**. This needs to be a one-page, short statement of the company's formal quality policy or mission.

A model statement is shown at the beginning of the model Quality Policy Manual in Appendix D. Words have been added to meet the new (9001:2000 version) requirements.

It must be signed by the Chief Executive of the company or the particular site and dated.

It is necessary to demonstrate that this has been communicated to, and understood by, all employees. Hence the language needs to be as *simple and straightforward as possible.* I was very impressed at one local factory (Vision Products, Pontyclun) where they employ some staff with disabilities, that the Quality Policy Statement had also been printed and displayed in Braille.

The easiest way to demonstrate that it has been communicated is to place copies (ideally within glass frames, to stop graffiti) at strategically situated locations in the offices and workshops. However, depending on the mentality of your workforce you may have to chose your display site carefully to prevent stickers, graffiti or other comments.

I have also seen it effectively communicated internally by printing on the top or reverse of internal job cards. I have seen it effectively communicated externally on the reverse of business cards.

Quality Policy Manual

Below the Quality Policy Statement comes the vital Quality Policy Manual (often referred to simply as the **Quality Manual**).

Note It is recommended that you may freely supply your Quality Policy Manual, possibly with the list of operating procedures, to potential customers. Never provide your actual procedures. They are how you run your business and must remain **strictly company confidential**.

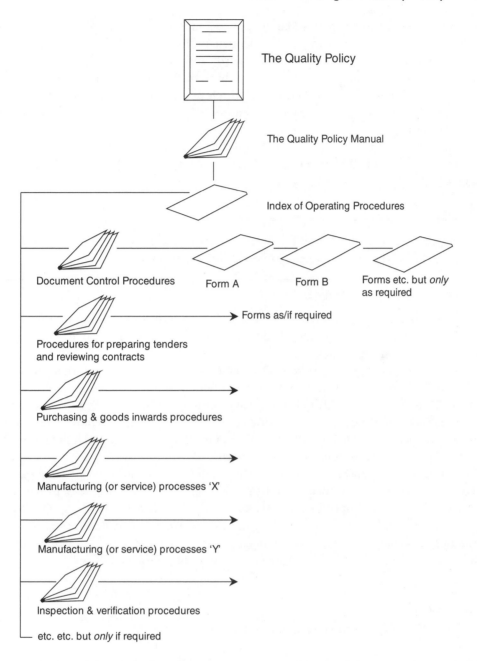

The Quality Policy

The Quality Policy Manual

Index of Operating Procedures

Document Control Procedures

Form A

Form B

Forms etc. but *only* as required

Forms as/if required

Procedures for preparing tenders and reviewing contracts

Purchasing & goods inwards procedures

Manufacturing (or service) processes 'X'

Manufacturing (or service) processes 'Y'

Inspection & verification procedures

etc. etc. but *only* if required

Fig. 5.1 Quality policy documentation

The purpose of the Quality Policy Manual is to:

(a) Translate the jargonized, stale and technical words of ISO 9001:2000 into more readable, easily understood language or words, but which also reflect the needs of your particular business. In effect, it becomes ISO 9000 for your business.

(b) Present an essential interface between your own very practical in-house procedures and the special requirements of an outside assessor.

(c) It states your company policy to each clause of the standard and hence identifies any procedures required. In effect it provides the essential foundation on which you build your procedures.

(d) It is a document that you can provide to clients on request in support of a tender or bid, or as promotional material.

Note You sometimes find QA Consultants stating that they need to provide training courses on understanding ISO 9000 for your staff. If the Quality Policy Manual has been worded correctly it becomes, in effect, ISO 9000 for your company. It is only necessary for your managers to understand your Quality Policy Manual. It is therefore NOT necessary for them to have an understanding of the details of the standard.

The accompanying model Quality Manual provides an ideal model for the small to medium-sized company. It will almost certainly be adequate for firms of less than 40–50 staff. However, if the product or service is straightforward and also the staff have not become departmentalised in their attitudes e.g. they think of themselves as members of the same company team, rather than members of and with loyalty to a particular department, the draft model Quality Manual and associated procedures will work for much larger companies, say up to 250 staff.

Please now act as your own QA consultant. Please go through the model Quality Policy Manual. Simply modify the model in red ink for your typist to draft your own Quality Policy Manual.

The Quality Policy Manual and procedures are in hard copy and on the following website: http://books.elsevier.com/manuals/0750649496

In several places you may see items in brackets, with '**OR**' or several brackets with '**OR**' , one after the other. The items in brackets are possible alternatives for you to consider if more appropriate to your own organisation.

When drafting out the Quality Policy Manual remember it is policy i.e. what you are going to do, not how you do it. 'How you do it' is contained within the Operating Procedures.

Also, throughout the Quality Policy Manual you will see (**OR** Standing or Temporary Work Instruction). Leave these in your first draft to check and probably remove later. The ideal situation is to have a two-tier system, e.g. Quality Policy Manual (including the Quality Policy), Operating Procedures (with proformas).

However, with a larger firm, or a small firm with lots of awkward little jobs that specifically need describing or lots of inspection, it becomes essential to introduce Work Instructions for that particular job or inspection. This is known as a three-tier system, e.g. Quality Policy Manual, Operating Procedures (with Proformas) and Work Instructions (with Proformas).

It may not become clear, whether you need, or you would prefer, to have Work Instructions until you document your Process Control and Inspection Procedures.

Choice of Wording
Do be careful with the simple meaning of the following words:

shall – policy is to do it every time
will – policy is to do it every time
may – this means optional
should – ought really to do it, but not mandatory (a word to avoid in the Quality Policy Manual or the Operating Procedures)
could – this means optional
instructions – to be followed in every detail
guidelines – used as guides, typically for trained staff, may be adapted, if appropriate

Controlled documentation
A controlled document has three things printed on it:

(a) a unique identity, either a description or reference number.
(b) if several pages, the pages have to be identified (typically 1 of 5, 2 of 5, 3 of 5, etc).
(c) date of issue or it may have an issue reference. It is not really essential to have both, but they are commonly both provided to satisfy the possible needs of a pedantic assessor.

A controlled document will also be **formally issued** (e.g. authorised for issue) and the distribution controlled in such a way as to ensure that all working copies in an organisation are of the same issue.

Occasionally you see attempts to combine the Quality Policy Manual and the Operating Procedures. *This is not recommended* for the following reasons:

(1) The target audience of the Quality Policy Manual and the Operating Procedures are usually very different. The Quality Policy Manual is mainly addressed to outsiders, often the QA specialists. The Operating Procedures are clearly aimed at the inside working operatives and the language and words need to be as simple and as straightforward as possible. Hence, if you attempt to put them together they do not read very clearly, especially as you would be attempting to mix abstract policy with practical pragmatic instructions.

(2) If you combine, you would no longer be able to freely supply your Quality Policy Manual in aid of a tender or bid to a prospective customer.

Having given you these few words of warning, now use Attachment D to draft your own Quality Policy Manual.

As you already know your business better than an outsider, it probably will not take you very long to produce such a draft Quality Policy Manual but would probably take a good QA Consultant at least three days, one day to visit you, one day to write/amend and then probably another visit to check and correct. By doing this first step yourself you will probably save between £600 and £1,200.

Details of the Quality Policy Manual

The following additional notes are to help you address particular features of the Quality Policy Manual.

Front page. At first drafting do not be concerned about issue/date, copy number, control or uncontrolled. This can be understood and completed later when drafting the Operating Procedure for Document Control.

Contents. It is useful to have the clauses in the Quality Policy Manual identical to those in the ISO 9001:2000 standard. It makes it very easy for the assessor to check.

Issue and Amendment Control Sheet. This one sheet controls the issue of the Quality Policy Manual in total. Similar content sheets appear in front of each of the individual Operating Procedures. This implies that if you change one item on one page *you re-issue the whole Quality Policy Manual.* (Similarly you re-issue a whole procedure). This may seem wasteful. However, please be assured that this is the simplest method and the path of least risk. The alternative sometimes used with separate issue of individual pages (or in some cases 'issues' for major change of individual pages and 'revision' or 'revs' for minor changes) can become a bureaucratic nightmare. Also, if you get them wrong, they provide very easy pickings for a pedantic assessor.

Introduction to the Company. This gives you the opportunity to present and sell your company. However it needs to be factual.

It can also be used to positively declare your scope or more usefully be deliberately **used to exclude certain items from your scope and/or the assessor's target area**. Suitable examples that could be documented and included are:

- 'We do not design, specify or draw any product or service. The 'design' or specification is always given by the client and its 'design' is therefore excluded from our Documented Management System and our scope of registration.'.
- 'We do not design, specify or draw any product or service. We do have to make sketches to provide guidance for our staff. These are identified, authorized and dated and filed in the project file after use. However, the "design" or specification is always given by the client and is therefore excluded from our Documented Management System and our scope of registration.'

- 'We do not design, specify or draw any product or service. The "design" or specification of our product has been well established and documented in textbooks etc., therefore "Design" is excluded from our Documented Management System and our scope of registration.'
- 'The scope of registration is for the manufacture and installation of PVC windows and doors. The wooden doors are individually crafted items and are excluded from the scope of registration.'
- 'Proprietary software is supplied, installed and configured to meet the requirements of customer's systems, hardware and ancillaries. Software design and development is therefore excluded from the scope of registration.'
- 'Included in our Operating Procedures are accounting, job costing, Health and Safety and Environmental procedures, which are included to make the Operating Procedures into an overall Management Control System (**OR** Company Wide Quality System). The above procedures are not required to meet our contract requirements or are an auditable part of our ISO 9000 scope of registration.'

Quality policy. Can be made more comprehensive if you desire. However, it needs to be understood by all levels of your organisation. It therefore makes sense to keep it simple.

Additional notes to bear in mind on some clause when drafting your DQS.

Control of Documents (Clause 4.2.3)

The method shown is the easiest and simplest form of control. That is when you change a page of the Quality Manual or a page of an Operating Procedure it is easier and less likely to go wrong if you change the whole document. Overcomplicated issue documents, transmittal notes, associated records, etc. can often provide rich pickings for a pedantic assessor.

Quality Planning (Clause 5.4.2)

Note, it is **not** a requirement of the standard that you have formal quality plans for all your products or services.

You may, if you have a complicated product, control your processes by using a quality plan for each product or service, if you so wish. However the use of quality plans across a company is relatively rare. Examples where they are used are, Engineering Manufacturing where a company may have various complex models made up of various part numbers which have to be assembled, tested and finished in a particular way. Also the Construction Industry, with many operations to control,

Responsibility & Authority (Clause 5.5.1)

The duties shown are obviously not going to be right for your particular organisation. However, they will give you a clear guide on what to include.

It is generally not acceptable to refer to a separate confidential job description. The purpose of publishing the responsibility and authority in the Quality Policy Manual is that everybody and anybody can clearly identify who is responsible for what.

As a general point it is worth noting that you can delegate authority and also the performance of a duty or for the execution of a task. However, you should never attempt to delegate responsibility.

It might surprise you to learn that the majority of firms now being registered do **not** have a Quality Manager or a Quality Assurance Manager. The duties shown for the quality manager are mandatory to meet the requirements of the standard. So these must be placed under the duly delegated 'Executive or member of Top Management who is personnally responsible for quality'. In a small organisation these duties may have to be listed under the managing director as his responsibility.

Management Review (Clause 5.6)

Failing to meet the requirements of this section are a common major non-compliance and hence non-recommendation for registration.

The items listed in the associated procedures **must** be reviewed with actions and conclusions recorded as a minimum at least annually. Most certification bodies would also expect at least one completed management review, with actions closed out, before the formal assessment.

I strongly recommend that you have monthly (or bi-monthly) management review meetings where you formally discuss, communicate and record ALL the items vital for your company e.g.

- work in progress
- future orders
- sales and marketing
- financial performance
- personnel issues
- quality assurance items.

The **quality issues** will then be discussed as **part of your normal management review**.

I have found, in practice, that if you try and call meetings just to review only the quality items, everybody seems to find an excuse to be elsewhere!

Design & Development (Clause 7.3)

This section of the draft Quality Policy Manual has been written to meet the needs of 90% of companies who do no design.

If you do carry out 'design' you must now under the additional requirements of ISO 9001:2000 be assessed and be registered for design, simply add the clause headings from section 7.3 of ISO 9001 and in practical simple language state what you are going to do to meet each clause. Note, in practice at least one design review is essential for each project.

To provide a model design procedure is not practical as they can vary considerably. However, they fall neatly into two types. These are explained in the additional guidelines in Appendix F, at the end of the model Operating Procedures.

Verification of Purchased product (Clause 7.4.3)

Please read this paragraph of the draft Quality Policy Manual very carefully to grasp its meaning. The meaning of this clause is often misunderstood and it often appears incorrectly as being ONLY goods inward inspection.

Identification and traceability (Clause 7.5.3)

Read the draft Quality Policy Manual and draft Operating Procedure carefully. This has been written specifically to provide a workable and practical situation for a typical manufacturing shop or service or distribution outlet.

Do not make the mistake of stating that everything will be labelled. This is not practical and will provide an assessor with innumerable non-compliances.

Also do not provide traceability unless it is a contract requirement or after great thought, you believe it to be worthwhile. It is usually very expensive. **A common mistake is to read into the standard that traceability will always be provided** to say the original suppliers' material or test certificate, or alternatively to put work into traceable lots or batches, etc. This is **not** required unless it is a contract requirement.

Customer Property (Clause 7.5.4)

Often referred to as 'free issue'.

Again a very commonly misunderstood clause of the standard. This requirement for this clause originally arose from the MoD/AQAP standard when the meaning was very clear and obvious. (e.g. within the same group in the factory would be empty 1000 lb bodies belonging to the RAF waiting to be filled with explosive. At the same time there would be identical empty 1000 lb bomb bodies belonging to the factory to satisfy overseas orders. As you can imagine the MoD would require stringent controls and safeguards in this common situation.

They would be worried that if the factory had problems, e.g. casting porosity, in their bodies, some enterprising individual might be tempted to do a swap or exchange.)

Be very careful before making a bold statement excuding this item and declaring '.............. Ltd has no Customer Property, hence clause 7.5.4 does not apply', in practice most firms have 'customer property' or material, or commercial-in-confidence information, e.g.

- An engine manufacturer provides crankshafts to a specialist grinding company to finish. These crankshafts are therefore 'free issue' or 'customer supplied material' to the grinding company.
- A large chain store selling its own brand of bikinis. It supplies the manufacturer with the cloth on a cut, trim and sew contract. The cloth is 'free issue'. It is important that any reject, excess, or material left over, is returned to the customer. (e.g. Not to be used to make a little private run of wrap around skirts, as witnessed by the author on one assessment.)
- The clothes or any items belonging to residents in a nursing home are 'customer property'.
- A photocopier supplier stores client photocopiers whilst a client (or potential client) trials a potential replacement machine. It is rather embarrassing if the client asks for his old machine back and somebody has 'borrowed' the drum for a spare for an emergency breakdown. Even worse if the stored machine has been sold and installed elsewhere. (Both of these examples have personally been witnessed by the author.)
- With some industries e.g. garages doing service and repair, breakdown recovery, the whole of the work is done on customer property i.e. the customer's vehicle. Similarly, in a Crematorium the coffin and contents are obviously customer property. However, there can also be very important subsidiary customer supplied material that may also need very carefully control, e.g. tapes or CDs for the service, caskets, flowers, names of attendees etc.
- If you were running training courses for an organisation, they might provide accommodation, food or transport as 'free-issue'. These need careful control to demonstrate they are not misused or abused.
- A firm is making trousers for several competing chain stores on sub-contract. They must take appropriate steps to ensure that a representative from one customer on a factory visit, is NOT allowed to see the designs or specifications of a competitor.
- A builder is given copies of architect's drawings for a project. He should not use these himself or provide copies to somebody else for a different project. The 'design' is the property of either the architect or possibly the customer.

Control of non-conforming product (Clause 8.3)

Do not make the common mistake of stating without thought, that non-conforming product will always be placed into a designated quarantine area. With large items (or large quantities of items) this may simply not be practical.

If you do have a quarantine area do not make it a fixed size by painting lines, etc. Invariably you will overflow the area on occasions, or it will be empty and somebody is desperate for space and then you've got a non-compliance or a mistake.

If you do have a quarantine area make it flexible in size by rope or hurdle barriers that can be moved.

Annex A of the Quality Manual – Showing the management structure

As a general guide such an organisational structure would not include names. So if anybody leaves, etc., it is not necessary to change the Quality Policy Manual.

However, in some smaller firms where one is promoting a 'personal service', it may be appropriate to include the names of staff.

I have seen consultants state that you **must not put names**. If you wish you can certainly put names, the only danger is that you will have to do the appropriate changes to the DQS when somebody leaves or is promoted, etc.

Chapter 6

Notes on writing the Operating Procedures

Why Operating Procedures?

You will note I have called the procedures 'Operating Procedures', not 'Quality Procedures'. This is deliberate. If you call them 'Quality Procedures' staff will think that they are in addition (probably in their mind an unnecessary addition) over and above their normal working practices, possibly only to be worked to during or just before a customer or certification body audit. Or, worse still, they will perceive that they are only the Quality Manager's Procedures or only for the Inspection Department and nothing to do with them at all!

It is therefore essential to call them something similar to the 'Operating Procedures' to show quite clearly that the procedures apply to **everybody** and this is how we all run our business, each and every day.

I have, and you should also, tried to put them into some sort of general natural order of the flow of work from start to completion (e.g. the processes to meet your 'product realization', in the heavy words of the ISO 9001:2000), through your organisation e.g. planning & tender, contract, design (if appropriate), purchase, goods-in, process, etc.

The order you actually tackle writing and implementing them is up to you. Try and achieve early success with as little bother as possible. Remember to draft and implement no more than two at a time (three at the absolute maximum) and work to get those right before introducing new ones. See the realistic time-scale in Fig. 4.1, Chapter 4.

The model Operating Procedures

The Operating Procedures shown in the accompanying model Operating Procedures are general working models. They almost certainly will not be right for your business without amendment.

Also do note the 'models' probably contain a lot more than you require e.g. they have several options or alternatives built in with (**OR** . . . . .).

How are you going to amend them? How will you know that they will be what ISO 9000 and the assessor wants or demands?

> This is one of the great misconceptions about ISO 9000, that in some way ISO 9000 prescribes that you must do this in a particular way and you must do that, etc.

It is often a common excuse for not introducing ISO 9000 that the requirements are too onerous, with claims that ISO 9000 dictates that we will have to label or number every component in the factory, that ISO 9000 means that every job must be written in a particular way with this amount of detail, that all staff will have to go on training courses and additional operations will become mandatory, etc. **This commonly held view that ISO 9000 dictates how to run your business is absolute nonsense**.

Using the model procedures provided *you decide* and document how you are going *to run your own business*. All you must ensure is that the Operating Procedures for your organisation satisfy the following rules:

- The Operating Procedures provide controls and demonstrate that you consistently meet your customers' specified requirements.
- If you are selling your own proprietary product/service, that you have written and controlled internal specifications for the products/service and you then ensure and can demonstrate that you consistently meet that specification.
- That all of your staff, who affect the quality of your product or service, are adequately trained, are competent and have access to those Operating Procedures affecting their own particular job. They understand and are working *exactly* to them.
- That the Operating Procedures that you have documented and implemented satisfy the general **policy requirements** of the various clauses of the standard, i.e. the standard says what controls your documented system should provide, *it does NOT say how you are to achieve them.*

Also remember the golden rule:

'Document what you do, do exactly what you have documented.'

As you prescribe documented controls a simple checklist of five 'W's and an 'H' is invaluable to ensure all items have been included in the Operating Procedures e.g.:

'I keep six honest serving-men (they taught me all I know). Their names are **What** and **Why** and **When** and **How** and **Where** and **Who**.'

Kipling

Notes on the authorisation and amendment record sheet

The purpose of this sheet is clearly explained in the notes in this chapter and the model procedure for 'Document Control'.

You will note in the models that I have put in several fictitious issues. This is to emphasise the point. It does not matter how many times you re-issue them. The assessors are more likely to be far more impressed if some procedures are at issue 6 or 8 rather than all at issue 1. In fact all at issue 1 may raise suspicions in their minds whether this is in fact a developed working system.

The vital thing is that the operating procedures explain exactly what you do in language that is clearly **understood by the least able member** of the team covered by that procedure.

You will note I have only shown three headings:

- purpose and scope
- responsibility
- implementation.

I believe that this is a sensible minimum and an adequate number of headings for most ISO 9000 procedures. You may have seen some laid out very formally e.g.

- requirements
- purpose
- scope
- references
- documentation
- definitions
- responsibilities

- implementation
- charts
- proformas, etc.
- records.

If presented too formally they can become unfriendly and extremely intimidating to your staff and operatives, which is not the object of the exercise. The objective is that the Operating Procedures become **their** procedures, how they do their jobs, etc.

To present such formal procedures within small firms can also become ridiculous. There is just no point throughout each and every procedure in writing:

(vi) References: ISO 9001:2000 Clause Number . . . .

Operating Procedure 1

Contract review

This procedure envisages a small operation of 1 to 30 staff, operating with a central and easily accessible **Enquiry and Contract Record Book**. It also shows a sensible two-tier system of reviewing contracts, or 'checking the orders before the process starts'. The values in £s are fictitious, obviously you amend them to suit your business, ability of staff, etc.

This one-book system is a very simple way of achieving the necessary controls. If this is not possible in your organisation, do not worry. Amend this procedure and put in the additional controls that you want. You may use a standard proforma similar to that shown for every contract, or you may place a rubber stamp mark on the contract (signed and dated) for smaller orders. Basically all you have to demonstrate and record is the items shown in paragraph 1(b) of the model procedure.

If appropriate to your business you may wish to design or use your own existing tender estimating forms. If you design it well you can use it to very effectively demonstrate 'Quality Planning'.

What if I supply mainly proprietary items, that rarely change from contract to contract?

Basically in the Operating Procedure state just that. Something on the lines:

'. . . . . plc manufactures its own proprietary products. The contract review consists of the Order Clerk (**OR** . . . .) reviewing the contracts (**OR** order forms) to ensure that the order is for one of our standard proprietary ranges of product. If confirmed that this is the case the order is signed and dated by the Order Clerk (**OR** . . . .). If the order appears not to be standard the Order Clerk (**OR** . . . .) raises the Non-standard Product Request Form, see Appendix . . . to this procedure. This is sent to the . . . . to see if this can be supplied and is to be accepted or negotiated.'

At this stage, if appropriate, the review can check for quantities required in a pack, special packs, requirement for 'own branding', special or additional tests, applicable discounts, credit rating, transport arrangements, colours, labels with bar codes, etc.

Bulk/annual order

Another common feature is that of a bulk or annual order with call-offs as required. If this is the case, just draft a simple two-tier contract review: a detailed contract review when main contract is placed, with limited confirmatory review at each call-off.

Bought in and supplied on (factored product)

What is item 4 in this Procedure? This is included to satisfy the famous clause 6.5 of BSI's regulations (or should I say infamous) previously known as Clause 5(f), for items or materials 'bought in and supplied on'. This is likely to be in all other UKAS certification bodies' regulations. This is to prevent a scenario as follows:

Factory 'A' gets registered to manufacture products X, Y and Z. 'A' finds it can be provided with a low grade substitute from a source in the Far East. 'A' purchases products from this substandard overseas source already boxed and with 'A''s brand or logo on the box. This is sold to the unsuspecting customers who believe that these have been made under the BSI/NQA/LRQA or URS registered 'or approved' QA system and controls at factory 'A'.

Note that this does not mean you cannot source **raw material** and **components** and properly specify and implement verification controls at goods inward and processing, etc. It means that it is simply not on to mislead people into thinking a product has been produced under an ISO registered system when it has been **totally** produced elsewhere.

I do not intend to imply that all products from the Far East are substandard. They are often that way because that is what the European (or American) customer has formally specified or implied by setting a low price. I was amazed at the high grade and consistently high quality standards produced in some factories in the Peoples Republic of China during my visit and assessment of some of their factories. Recently, more of these good factories have themselves become ISO 9000 registered as it is an international standard.

Recording minutes of meetings or site visits (Para. 5 in model procedure)

Applicable to installers, site maintenance, etc. Probably not applicable to most manufacturing companies and could be discarded.

Contract review on verbal orders

Verbal orders are not an ideal situation, as the standard when first introduced envisaged a factory making 'widgets' to a written customer order, accompanied by a specification and/or engineering drawings. Verbal orders are also not an ideal situation from a common sense, business point of view. With verbal orders you are exposed to risk that whatever you produce can always be rejected as 'that is not what we ordered'.

However, if the real-life situation or market you are trading in means that customer orders are not written, state exactly that that is the situation in the procedures. The methods of dealing with verbal orders and documenting them are as follows, in order of preference:

(1) Send your understanding of the contract by return by facsimile machine. The added beauty of the fax is that the receiver's fax number appears on the print-out and it is difficult to claim they haven't received it.

(2) Send your understanding of the contract by post. Send by recorded delivery if the value of the contract is high. It is always advisable to send the above if you get 'an instruction to proceed' in advance of receiving the contract. I believe under UK Law you do not have a formal contract until you confirm this instruction. In addition you could proceed and then find that the written contract, when it arrives, is not to your liking.

(3) Write down your telephone or verbal order in your Enquiry and Contract Book and 'read back' the contract to the sender. If the order is made up of part numbers with descriptions a simple but very effective cross-check can be built in e.g. the customer reads the part number to you, you read the description of the item or part back (or vice-versa).

(4) At trade-counters the contract review could be a 'read back' of the order with the customer signing a copy of the invoice or picking list before collection from the warehouse shelves. This works very well with modern computerised warehouse stores.

(5) The last option is a straightforward 'read back', not that effective but the best one can do. It is common in retail outlets or restaurants or fast food outlets: note the careful 'read back' next time you are in a modern fast food outlet.

Operating Procedure 2

Purchasing

Basically all the standard is asking is that you know exactly what you are buying and you know who you are buying it from.

Approval of suppliers

The standard does not say an approved suppliers list is essential but I have never seen another way of effectively addressing these controls. In theory it will be possible to state '. . . . plc have only three suppliers being (1) . . . . (2) . . . . (3) . . . . with whom we have dealt for 'X' years. We do not intend to purchase elsewhere.' However, this is unlikely to be a very common occurrence.

Please do not bring the QA process into disrepute by sending QA questionnaires to all your existing suppliers. Please go through the sensible stages of:

(i)   Identifying and listing ONLY the suppliers that affect the quality of your product/ services.

(ii) Establish if they are already ISO 9001:2000 registered (or 9001/2:1994 up to 15th December 2003). Telephone or fax them, get a copy of their certificate, check it's from an accredited body and their approval is the same as the goods/services you require.
Beware 'presentation' certificates; you want a copy giving the scope of approval.

(iii) Have you factual objective evidence of buying from an existing supplier with purchase orders and evidence, over a significant period of time, of good delivery? Can you document these and show evidence of evaluation? See form in Appendix E.

(iv) If (i) to (iii) do not apply, only then send questionnaires/audit visit, etc. The questionnaire in Appendix E (Operating Procedure 2, Annex C) as an example may seem rather long winded. However if you send a shorter one, (an example of a one page questionnaire also provided (Annex C2)) people who have not introduced QA will be completely bewildered. From the replies you get on my 'explain it to them' example it will be pretty obvious that some have not got a clue.

You should be aware that this part of the Purchasing Procedure can take some time to get fully implemented, possibly 2 or 3 months if you have a large number of suppliers.

However, if you prefer, there is nothing to prevent you from splitting this procedure (or any other procedure) into three, e.g. 'Procedure for approving and controlling suppliers', 'Procedure for purchasing' and 'Procedure for Goods Inward Inspection'. Do remember to ensure your cross reference with the Quality Manual remains correct and also be aware that it exposes you to more risk as it gives you more procedures to control. However it can make the individual procedures shorter and easier to understand.

Purchasing procedure

Should be straightforward and in the majority of cases documenting what you probably already do.

However, be aware that if not fully implemented, purchasing can give rise to a failure at assessment by providing a multitude of non-compliances. It is so easy to check. The Assessor has only to request your Purchase Orders for the previous three months and go through them and then in a very obvious fashion it can start to fall apart e.g.

- Purchase orders that have been placed with no restrictions/verification on to unapproved suppliers.
- Verbal orders, without paperwork/confirmation.
- Illegible purchase orders.
- Orders for 'three of what we had last time'.
- Non specific orders e.g. 4 dozen, 25 mm × 5 mm diameter bolts i.e. not specifying material, stainless steel or plastic, hexagon head, thread type, etc.
- Unsigned purchase orders.

Goods inward inspection

Should be straightforward to select the appropriate methods for your organisation from the examples in the model procedure. However, I would strongly recommend using at least a 'Goods Inward Record Book' for **all** organisations. Try it and you will probably find that in six months you have built up an extremely valuable record to which you will be referring to for critical information that is immediately available.

> Note also that the goods inward inspection **must** be against a copy of **your** purchase order; not against the firm's advice or delivery note. Otherwise you, or your staff, are only checking that the supplier has delivered to you what they wanted to deliver, which may not be what you ordered! Stationery suppliers, in particular, and other stockists will often deliver the nearest stock item or alternative brand item that they have on their shelves.

Also, if goods are delivered to working or construction sites it is **essential** that the responsible person on that site has a copy of your purchase order to check, sign off and return to the office.

Note that I have written out **vendor rating systems**. You often find these put out to small firms with only a few suppliers, by 'Quality Consultants'. Vendor rating systems only apply sensibly to large firms with many suppliers.

Operating Procedure 3

Process control and in-process inspection

You will find detailed within the models various proven simple methods of controlling and inspecting your processes.

Read through the various options and choose the one most suited to your organisation and use that model to document *how you manage and control your business.*

> It is not generally appreciated that your procedures reflect how you have **organised the management** of your business. It is **not** necessarily product-related at all, so **don't make the classic mistake** of just copying the procedure from somebody making a similar product.

Also dip and pick into the other model process control procedures if there are particular items, controls or forms that are useful which can be modified.

This procedure on 'Process control and in-process inspection' is one where it may be appropriate to add these two extra main paragraph headings:

References

If there are International, National (e.g. British, American, Japanese, etc.) Standards, or trade, customer or internal specifications, applicable to your product these should be listed.

Definitions and abbreviations

If your industry or organisation has particular names, terms or descriptions that would not be generally recognised by an outside assessor or a new starter to your factory, list these, too. If these become lengthy you may wish to show the definitions as an Appendix or as a handbook if particularly complicated or interesting.

> **Note** I have shown in the model procedures that the responsibility for the 'Process Control Procedure' should be the production director/manager (i.e. it should **not** be shown as the quality manager)

Adapting the model

If there are several processes with different departments, types or systems or management controls or styles, change this Process Control procedure and substitute an appropriate number of **Departmental** Procedures, e.g.

Operating Procedure 3A for the Pickling Shop.
Operating Procedure 3B for the Sand and Moulding Shop.
Operating Procedure 3C for the Pattern makers.
Operating Procedure 3D for Melt and pour of metal.
etc, etc,.

The need for Departmental Procedures in this fashion would normally only occur in a larger organisation. It may also be appropriate and helpful in such Departmental Procedures to include an organisational tree for each department (as per Annex A of the Quality Policy Manual).

> If you do have to document procedures by department it is **essential** that each Departmental Procedure is authorised by the heads of departments (or better still they have drafted it). It is essential that it describes how they run their departments and when the organisation goes for assessment it is *their name in the frame for non-compliances* against their department and their documented procedures.

I have shown departmental headings for a typical engineering organisation but it could just as easily have been a service scenario e.g. administration department, recruitment department, litigation department, public relations department, membership department, etc.

Inspection and verification

You will note that I have included the in-process inspection, final inspection and verification into the process procedure. The goods inward check is included with the Purchasing Procedure.

If you prefer and it seems more appropriate for your products, customers or management system, these can just as easily be put on their own in an 'Operating Procedure for Inspection and Verification'.

What is the difference between inspection and verification? Inspection is a detailed check on the product/service where verification is a separate independent overall check. It can also include the inspection of your inspectors i.e. who checks the checkers?

Operating Procedure 4

Control of materials

When making, installing or maintaining a tangible product
The labelling system prescribed in the model is for a manufacturing, or machine shop, or installation type of environment and should be straightforward and easy to follow. The key thing to remember is

> **not** to claim you will label everything. This would be a mistake as (a) the place finishes up looking ridiculous and rather like a Christmas Tree with green labels everywhere

and (b) you will finish up with innumerable non-compliances on assessment as inevitably individual items would be found unlabelled. So note that you label only **when necessary** to identify the product.

Similarly you should, if possible, devise an inspection status control that requires you only apply an inspection status label if required, i.e. **if it doesn't have an inspection status label on it stating 'hold', 'quarantine' or 'reject' it is acceptable material or work in progress.** Read carefully the Model Procedure reference the green or accept label (Operating Procedure 4, para 3.2[f]).

If you decide to affix green/accept labels to outgoing product, do ensure that the labels are completed carefully and very, very neatly. If they are carelessly scribbled on, it completely destroys confidence and the desired effect.

The inspection status labels shown in the model give a sensible measure of differentiation and control. It is almost universally used in one form or another and is simply based on the 'Traffic Lights' colour principle, e.g. Red = Stop, Amber = Stop/Caution, Green = Go. They can be cards or labels, or with something basic such as castings it could be a painted cross or ring of appropriate colour.

> To satisfy the standard you can get away with only one 'Reject' label. However, in practice it does not work very well

as suspect or doubtful materials are only 'rejected' and scrapped as a **last option**. They are usually **repaired**, **reworked**, offered for **concession**, **sifted**, etc. – see Model Operating Procedure 7. Hence under this one card 'reject' there will be items for 'scrap' and also items for 'rework' which is a sure recipe for confusion and will lead to mistakes.

Think carefully before you dismiss the 'special' label/card as unnecessary. I can remember vividly two incidents where this decision was reversed very quickly.

(i)   A Landscape Contractor thought it unnecessary until they found they had one important customer who had a Landscape Architect who insisted on going to the nursery and selecting individual shrubs or trees for a particular location.

(ii) One manufacturer of standard off the shelf fencing panels also thought it unnecessary to have this simple extra control, until they produced 20 'deluxe' fencing panels at enormous expense for the National Exhibition Centre trade show. They went to collect them before the exhibition but they had gone with the rest on the B & Q lorry. They then had to make do with their standard product for the exhibition. Some lucky individual now has the best and most expensive fence in the UK.

> Do not claim and introduce traceability if you do not require it.

A very common mistake, often made under the guidance of a 'consultant', is to introduce traceability as a requirement of the ISO 9000 standard when it is not required. To provide traceability with the necessary controls and records is very, very expensive. The other aspect to consider carefully before introducing traceability is that it needs very careful control. Once the chain of documenting evidence is broken it may be impossible to re-establish.

If you have a machine shop, read very carefully the model procedure for 'bar stock'. To be successful at an assessment such a system must be implemented fully.

> Be warned, poor control in a machining shop of metal bar identity and particularly the 'off-cuts' can be devastating on the assessment.

You must implement fully. If necessary document as a 'Standing Work Instruction' on its own. I strongly recommend that you, or someone specifically allocated for this task only, check the bar in the stores and the machine shop including the corners, lockers and 'doorstops' **the day before** and **early each morning**, during the formal assessment. It only takes one careless individual five minutes to saw two or three bars ignoring the controls and they could have created 7 or 8 non-compliances.

When providing a service without a tangible product
This can be quite a difficult part of the standard to address as the standard was originally written with a tangible product clearly in mind.

I have provided a model for you to build around. You may have to experiment with three or four simple systems to get it right for your service and in particular, how you have organised your business. The basic requirement is that in the 'processes', your documented system and controls enable you to show, or it must be self evident:

(a) what it is you are doing, and on what
(b) who is doing it (the records must show for future reference, who did it)
(c) when (usually on a daily and monthly basis is adequate).

Usually the most sensible way of addressing this is a bespoke shift or daily record sheet or log that identifies the tasks through the day, who did what, etc. which is completed as the day progresses.

It is clearly vital in businesses that are document-orientated e.g. recruiting agencies, consulting engineers, estate agents, solicitors, etc. that, after it has been booked in, each document or letter that is in the process, or a copy of any letter/document that has been sent, is clearly identified regarding:

* what project, contract or person it relates to;
* date;
* who is dealing with it or who dealt with it.

Note: the need to know 'how' you are doing it, is given under the Operating Procedure on Process Control.

Operating Procedure 5

Control of measuring and processing equipment

If you read through this procedure it should be easy to implement.

I strongly recommend that you send all gauges and instruments outside for calibration. In my experience it is simply not cost-effective to do them in-house particularly in a small company, and it is doubtful if it is really truly cost-effective even in a very large company. Also it is doubtful if you have the facilities, environmental conditions, equipment, expertise or staff to provide a calibration service in-house (see ISO 10012-1 and ISO 10012-2 if you want to get an idea of how onerous the requirements can be).

I have also provided the option of introducing a two-tier system of 'Inspection' and 'Production' gauges and instruments that can be introduced in some cases to reduce costs. Again, do cost it out carefully and honestly. I believe it is rarely cheaper to do these instrument checks in-house and is hardly worth the bother. You will find by going for competitive quotes that the prices charged for calibration are usually very reasonable; also many calibration houses provide a collection and delivery service with short turn-round times.

It should also be noted that

---

by placing the calibration outside you have substantially reduced your risk of failing assessment.

Just look at the length of this clause in the standard and the possibilities for non-compliances. Clause 7.6 is one of the longest clauses in the standard, with a whole 'larder' of possible non-compliances. Take my advice and subcontract this part of the system. It will:

* probably be more cost-effective
* reduce substantially the risk of non-recommendation at assessment
* allow you and your staff to concentrate on what you are best at, e.g. your own product/service.

If you are producing equipment that depends on critical or close tolerances, note carefully the option provided in the last paragraph of 3.8 in the Model Operating Procedure. You need to be able to show that you have controls that allow you to take effective actions if a gauge or instrument is found out of calibration when on its routine check and therefore may have given an incorrect reading on critical parts. See penultimate paragraph in clause 7.6 of the Standard. How can comply with this unless you know what instrument has been used on what batch? Check your controls. Can you show that these critical parts are only tested on this one instrument? Or, check your record keeping as you may need an additional record card/form for these critical measurements with an additional column for the instrument identity/number.

You will note that I have taken rules and tapes out of the calibration system. This is sensible as they are used for coarse 'eyeball' checking. This is accepted by most certification bodies and assessors. If you have one that tries to insist that you calibrate rules and tapes, fight very strongly against doing it; it doesn't make sense (are all your employees' eyesights calibrated!) and it is very expensive.

If necessary, compromise on a check system of identifying the rules and tapes and calling them in (say six-monthly) to check visually for wear, chips, damage, etc.

If the assessor still insists on calibration of rules or tapes you may wish to consider changing certification bodies.

If you do *not* have any *gauges* or *inspection equipment*, state so as an exclusion in the Quality Manual and do not have a calibration procedure. Do *not* unnecessarily purchase one micrometer just for the assessor's benefit. Do *not* start a system to 'calibrate' people, unless it is a contract requirement.

Operating Procedure 6

Training

Very straightforward usually. Many firms already have suitable training record cards, etc. and it is often simply a case of documenting what you are doing already.

Strictly speaking, to meet the requirements of the standard the training record/appraisal need only take place on staff/employees whose actions affect the quality of product/service. It usually makes more sense to encompass everybody in this part of the system.

Operating Procedure 7

Management review, internal audits and corrective/preventive actions

This procedure is at the very centre of implementing ISO 9000 effectively and obtaining the full benefits.

Also, its actual implementation makes it very clear to an experienced external assessor (and also even clearer to your own staff) whether a firm's management is really committed to improving their quality or just paying lip service in order to get the certificate on the wall.

Management review

It is not stated in the standard, but it is generally accepted as a minimum that at least once per year there is a formal (e.g. documented, dated and signed) management review of the QA System and the associated activities and records.

Whilst this will satisfy the assessor I strongly believe this minimum approach becomes something of a farce. Annually is obviously far too long a period to leave critical problems unattended. Also if you try and hold such a meeting with the other department heads expect second-rate substitute attendees.

I strongly recommend a nominal monthly or bi-monthly total (general) management review covering typically the topics shown in 3.1 of the model procedure. This then becomes a very meaningful management review with the quality aspects as a totally integral part of running your business.

Although I recommend monthly or no less than bi-monthly management reviews, you will note that I have documented at least every three months in the model procedure. This is so that you don't write yourself a non-compliance. If you were to say you are going to hold management reviews monthly (or bi-monthly) that is **exactly** what you must do.

Even if you are the managing director of a very small firm I would strongly recommend a monthly review or written report on the lines of the topics shown. This discipline will provide a check list of items you should have done and set yourself targets for next month. Also in a few months time you may realise that this becomes an invaluable historic record that is referred to continually e.g. when did . . . . have an accident, was it last February, March or April? When did we have the last large order from . . . .? How does our production volume and quality performance compare with this month last year? Such a disciplined management review can also be very convincing evidence of a well-run business when presented to the Bank Manager or the local Enterprise Agency when looking for additional funds.

To satisfy the standard they must be chaired (or written in case of very small firms) by the managing director (e.g. Top Management). Also to work effectively they must not only be chaired by the executive, but the executive must be obviously committed to the system and looking for actions to be completed effectively in an appropriate time-scale with comprehensive reports, etc.

Internal audits

The system shown in the model procedures has now become almost standardised and is typical of those introduced and successfully registered.

You will note I have written the option in of using an outside assessor (consultant, colleague or friend) to do the internal audits. As long as they have appropriate qualifications and experience this will be acceptable. However, **do remember to put them on the approved suppliers list!**

Be aware that the easiest way to carry out an Internal audit of a company is by going through an annual schedule of checking the firm's own procedures in turn and possibly by checking departments or areas or sites as appropriate.

You do see some internal audits scheduled against the 'clauses of the standard' mimicking the approach of a trained external lead assessor. I would recommend that you do **not** take this approach on your own internal audits. It will create confusion, conflict and possibly despair as you get different inexperienced internal auditors trying to re-interpret the 'words' of the standard for your business.

Also do not audit yourself to death!

It is quite common to find that QA Managers in some medium-sized organisations have **written themselves into a permanent job** by documenting an 'over-the-top' audit regime.

Generally to satisfy the assessor you need only show that you are effectively auditing each portion of your documented system at least once per year. However you may have to audit some again if you have lots of non-compliances found. Also if an area is obviously critical you may need to do it more than once per year.

(This is not stated in ISO 9000, but an annual audit cycle has become the generally established practice.) However, it should be noted that it will be necessary at the start, just prior to the formal assessment to have an intensive period of internal audits, see Fig. 4.1, Chapter 4.

It is essential that each procedure is audited and corrective actions completed prior to the formal ISO 9000 on-site assessment.

You will also see that I have built some flexibility into the model programme for the internal audits, allowing them to take place within a three-month period. Again do not write yourself non-compliances! If you schedule against specific months, that is what you must do e.g. Internal Audit schedule for April carried out in April no problem, Internal Audit scheduled for April carried out in June you have just given the external assessor an unnecessary and silly non-compliance to record against you. When doing your schedule do bear in mind items such as peak holiday periods, high output for any seasonal products, etc.

Audit checklists

Some organisations or consultants recommend that the auditor goes around with a pre-printed checklist. You may think this is desirable and they can easily be produced by asking the obvious questions: Can I see your copy of the Operating Procedures? Available/latest issue? Look at at least two contract orders and check for: evidence of formal contract review, signature, date; contract same as the offer; has it been amended, is there a subsequent review, etc.

These can be useful for an inexperienced or nervous assessor. However, do be warned that they must be revised or withdrawn after two audit cycles or it just becomes a stale routine and a facade. You may suddenly find you are getting non-compliances arising from an external assessor in areas where you have just completed an internal audit. This is because the internal auditor has not been given the opportunity to ask the obvious questions: 'How do you do that?' 'Why do you do that?' 'Why is it called that?' 'What happens if that member of staff is ill and absent?' 'Why is it called that?' etc.

> It is often the 'seemingly daft' or unpredicted question asked by a knowledgeable outsider, who does not know your processes or product/service, which opens a chain of questions that exposes the weaknesses of your management systems.

Corrective and preventive action

Note in the standard it is 'preventive' action and not 'preventative action'. I don't believe there is any difference and none of the Lead Assessors I'm associated with would even comment if you used the wrong word, but be aware of it.

The approach from a manufacturing system with product, to that of a service industry is manifestly different and this is reflected in the alternatives shown in the model procedures.

Your implementation of this section, as with your implementation of Management Review and Internal Audits, will clearly show your commitment to the QA process.

In the manufacturing system I have shown a model concession system. If it does not apply to your organisation write it out, e.g. '. . . . plc always produces items to drawing and specification. If defective work or product is found it is always replaced with good (**OR** unsatisfactory contracted site work is to be reworked to bring to specification). Hence a formal concession system is not appropriate to . . . . plc.'

> However, do be aware that a formal concession system is one of the most powerful tools available to a QA manager when up against a powerful 'get it out regardless' production director.

By involving the design manager and marketing manager in the process of minor concessions and forcing the major concession to be referred to the designer or the customer it can quickly bring the products back to specification. Do note that in this formal concession system the production director/manager is not involved and hence cannot pressurise or threaten the QA Manager.

Even if you don't have any customer complaints or haven't had any yet,

> do have the Customer Complaints Record Book ready and available at the time of the formal ISO 9000 assessment.

It does show that you have really implemented the system fully.

Operating Procedure 8

Document and data control

It should be relatively straightforward to modify the models to document and implement. My normal recommendation is that the QA manager is aware of its requirements but do not formally introduce document control until three months before the assessment.

However, if you judge you can implement fully from the start of the project then make this the first procedure to be issued.

Do be aware that, if not implemented fully in the last couple of months before the assessment, it can be the cause of a multitude of non-compliances and a possible non-recommendation at assessment.

I strongly recommend, that 2 weeks before the assessment you get somebody completely independent to check that **all the issues are exactly** the same.

These are really very stupid and unfortunate non-compliances to receive on an assessment, especially as they have nothing to do with how you really control your business. Now that you are aware of the dangers, just before the assessment, check them and remove the inconsistencies.

You will note that on controlling the pages I have given the alternative options of putting a date only to identify the issue, or the alternative of putting a date and an issue number. The second alternative has more risk of making a mistake and just makes it more complicated than necessary. However, I have shown Issue Number/Date as assessor's from one certification body (not URS) will tend to insist on an issue number.

An alternative method of page numbering is to number the pages with a '+' following the number if there is a following page and a 'minus' on the last page. Hence a five-page document would be numbered 1+, 2+, 3+, 4+, 5–. This method may have advantages if you anticipate frequent changes, as if you add pages it does not necessarily affect the intermediate pages.

The first option in Paragraph 3.6 of the model procedure specifies the easiest and simplest way of controlling the actual change. That is by the quality executive personally changing all the procedures (or the Quality Manual). By this simple device you remove a whole tier of outward and returned transmittal notes. Inconsistencies in these transmittal notes and also action or incorrect action by the receiving holders can cause a host of non-compliances.

Ensure that a copy of the Quality Policy Manual and Operating Procedures are available to all staff. If you have issued Work Instructions ensure that they are immediately available or displayed at the workstation.

Operating Procedure 9

Records

This is now (under 9001:2000) an essential procedure in itself. However, I always included previously, as I have found in practice that this provides an excellent summary and checklist of your records. It is very useful to check that you haven't forgotten any vital records and also to quickly demonstrate to an external assessor that the appropriate records are kept.

Note the requirement to address 'disposition' of records. This requirement of the standard is often omitted. Not particularly a vital omission in an engineering firm. Absolutely vital in a recruiting agency, doctor, solicitors, etc.

Also to consider is the repercussion of the Consumer Safety Legislation (Product Liability). If a product has a potential product liability implication (i.e. you can be sued because the product has injured someone) you probably need, in the UK, to keep your pertinent records of contract review, inspection and test, design review and tests, etc., for 15 years! You are liable as the manufacturer, or importer into the EEC, for any injuries 10 years after you put a product into circulation. The person suing you has 3 years to decide they have been injured and allow 2 years for the legal processes, hence 15 years.

Beware of records on heat or light sensitive paper such as faxes as they deteriorate rapidly. You may need to consider if as a standard procedure you should photocopy such items. If you have to keep records or drawings for this length of time it is safe to keep them on computer floppy disks. What is the life of a disk? Will a computer of that design be available in 15–20 years time to read your disk? Will the floppy disk drive, if they exist, be the same size?

What are the minimum records one has to keep to maintain ISO 9000 registration? Look through the standard:

Where it says '(see 4.2.4)' this gives you the clue; for each of these elements you **must** have adequate records.

Do not go over the top with **additional** record-taking. Two little guidance rules:

- If you are not going to, or haven't got the time to, analyse the records, don't bother taking them. Also, if you have too many records, you will get 'paralysis from analysis'!
- If having analysed them you do not intend, or cannot afford, to take any actions, don't bother taking the records.

Chapter 7
Notes on training internal auditors

It is a requirement of the ISO 9000 standard, clause 8.2.2 that you establish and maintain documented procedures for planning and implementing internal quality audits.

It also states that auditors should not audit their own work.

However, unfortunately it does not state in this clause that the auditors have to be trained, or stipulate what is considered to be adequate or appropriate training.

The requirement for training of auditors has been implied from two other clauses. That is, Clause 6.1, that you will provide adequate resources, and Clause 6.2.2, that you will provide competent personnel performing activities affecting quality. Hence, there is **no written statement of what is 'adequate' training** and it is a very unsatisfactory situation. Therefore, this places you in the situation where you must decide what is adequate training and the assessor has to demonstrate that it is not, in order to raise a non-compliance.

In truth, having documented the audit procedures, it is not that demanding to carry out the basic requirements to question and seek factual evidence and confirmation that staff are:

1. in possession of adequate procedures covering their work,
2. working exactly to the documented procedures,
3. working to the customer's, contractual or . . . . . plc's own specifications.

The ability to do this thoroughly, conscientiously and with tenacity is more a function of personal quality than that of training. Some people, by their very nature, do not have the courage to say 'Show me'. They are scared that somebody is going to say 'Are you calling me a liar?', or if they record a non-compliance the knee-jerk reaction may be, 'I'll get you back for that', etc.

It should be noted that it is **not** a requirement of an internal auditor to ask if the controls satisfy the 'words' of the clauses of ISO 9001:2000, etc.
However there is a new requirement in ISO 9001:2000, to ensure the management system continues to meet the requirements of ISO 9001:2000. This has been addressed in the model procedures.

The alternatives for training internal auditors, placed in order of preference to satisfy an external Lead Assessor are:

(a) Courses for 'Internal Auditing' run by the certification bodies themselves. These will be accepted without question by an IRCA Lead Assessor. They tend to be expensive (£300–550, 2002 figure) and also I believe give the attendees too much information as they are usually 'padded out' to spread over two days. I have seen firms send their auditors on the Lead Assessor Course which is very expensive and usually residential over five days. This is not necessary or recommended unless they are of graduate ability, as they will get completely confused. Also you as the employer may be funding your employee's next external career move!

(b)   Courses on 'Internal Auditing' by private consultants or colleges that have satisfied the requirements to become recognised by IRCA for the training requirements for Internal Auditors, will be accepted without question by an IRCA Lead Assessor. The IRCA has set up a Register of 'Internal Auditors' and also a rigorous scheme of assessing and approving of courses. Because of the relative high costs of administrating the register and assessing and auditing the courses, with the inevitable addition to the fee charged for such courses, it is difficult to assess whether this in time will become the accepted norm.

(c)   Course in 'Internal Auditing' run by either an **ISO 9001** registered QA Consultancy/ QA Trainer or a **Registered Lead** Assessor. This would almost certainly be accepted without question by an IRCA Lead Assessor. It will probably be of one day duration only and with a more realistic fee rate. Typically about £150 + VAT (2002 figure).

(d)   Having one 'trained' internal auditor who has been on course (a), (b), or (c) above, runs an internal course/seminar and escorts people whilst they are carrying out their first audits. If the details or notes of your 'course' are available and are sensible and also there is objective evidence of effective audits being carried out subsequently, it is highly likely that this will be accepted by the IRCA Lead Assessor.

(e)   Courses on 'Internal Auditing' run by a local college or unqualified or unregistered consultants. The Lead Assessor may look at the course notes supplied and will almost certainly look very closely at your internal audit procedures and records at great depth. In all probability you will satisfy the assessor but the contents of such courses are very variable. They are often run by professional 'trainers' looking to complete a course portfolio with notes 'lifted' from elsewhere. Not a recommended alternative.

I have not attempted to write a set of notes on 'How to do Internal Audits'. I believe it is one of those things that can only be taught and demonstrated.

> If you can afford it, one of the **best methods of introducing internal audits**, is to take a local QA consultant of good reputation who is also either ISO 9000 registered or a Registered Lead Assessor. Have him/her run a one-day course in-house on internal auditing, six to eight weeks before the formal assessment. Then as part of the course for an extra one or two day(s) s/he should carry out a complete schedule of internal audits with your trainees as witnesses. This is a very useful and effective way to use a qualified consultant.

In a couple of days it achieves several objectives:

1. Provides you immediately with records of internal audits for the assessment carried out by a qualified professional.
2. Confirms that your system meets the requirements of ISO 9000 and also *removes damaging non-compliances before the assessment.*
3. Have somebody independent look over your system, prior to assessment. The truth of the matter is, you don't see your own mistakes.
4. Provides the best training for your internal auditors to see a professional test the DQS to find what is not working effectively.
5. Provides training for the staff who are being audited. They have an experienced outsider quizzing them by saying 'You say you do this; show me', 'Let me see two more examples', 'What if . . . . .?' etc. and this is accepted because it is an outsider. When your internal auditors then ask similar searching questions in a few weeks time it is accepted as the norm. You don't get any of the adverse reactions that can arise.

For the small firm it may well be advantageous to simply subcontract the internal audits to an approved ISO registered QA consultancy or a registered assessor. As this provides a regular low risk income most good QA consultants would quote only 2–3 days per year for this service, at a low fee rate, including keeping your documents up to date and amending them as required, attending your management review meetings and telephone helpline.

Another alternative is to request the QA Manager of one of your larger customers or suppliers, if they are trained auditors. Most would jump at the opportunity.

This same arrangement applies to larger firms, however with larger firms you would lose the substantial added advantage of people from different departments auditing each other and it can create interdepartmental co-operation and communication.

The Managing Director should seriously consider being an Internal Auditor himself/herself.

It gives you a glorious and legitimate opportunity to witness all the activities in your organisation.

Chapter 8

The role of the Quality Assurance Manager and the relationship between ISO 9000 and practical TQM (Total Quality Management)

Before we consider roles, let us reflect on the various titles given to the practitioners asked to carry out ISO 9000 tasks and what these titles imply.

The titles of 'Quality Assurance Managers'

Quality Manager

This title implies that the title holder 'manages quality' and has divine powers or ability on their own to achieve quality in the product or service. This is in practice impossible for the Quality Manager to deliver. The provision of a product or service which meets agreed specifications is, in the main, **directly** related to the attitude and performance of other managers e.g. the Design Manager, Purchasing Manager or Production Manager.

Chief Inspector

This title appears to be a leftover from the 1970s but in fact goes back in history to the Pyramids, Great Wall of China, etc. This implies that quality can be inspected into the product by visual checking or testing and hence removing every single defect. Apart from the most archaic production manager/director this is now widely accepted as a nonsense. It can easily be demonstrated that 100% inspection does not remove all the rejects (see Chapter 3). Also that it is very expensive in person/power and the cost of scrap, rework and customer returns. Why then does the title still exist?

Quality Controller

This is a Chief Inspector function with the addition of a multitude of SPC (Statistical Process Control) or other quality control systems and charts.

QA Coordinator

This implies responsibility for Quality Assurance (QA) without any **executive authority** to make it happen. I often find individuals struggling with this title, having a very clear idea of what is required to be actioned but reporting to a committee, (steering group, quality council, etc.) to make the decisions. Unfortunately committees tend to promote the ideas of the strongest member, often the production or sales director. Also committees tend to compromise. With some QA topics, especially when going for registration or certification (or higher quality accolades) compromise is not always a viable alternative.

QA Representative

This title is often used to satisfy the perceived requirements of certification bodies. It tends to imply that this individual will deal with customer complaints or defective product or service. Also the QA Representative will 'bat' the questions of an assessor from the certification body. In between times the quality system can look after itself! In practice, the only time this approach can be legitimately applied is where the firm has formally subcontracted the practical maintenance and improvement of its QA system to qualified consultants, who are working on a part-time basis.

Quality Assurance Manager (Director), Quality Systems Manager or Quality Engineer

In the author's opinion these are currently the only legitimate titles for the individual who is given the responsibility and authority of documenting, controlling and implementing effective Quality Management Systems with associated auditing, product/service verification and bringing about continuous improvement.

The Role of the QA Manager

The idealistic role of the QA Manager is to address all the items or building blocks shown in Fig. 8.1 (or alternatively Fig. 8.2, if process control, health and safety and environmental issues are also included). Let us briefly consider each of these items in turn.

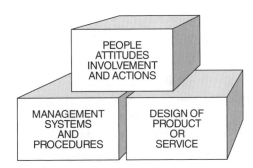

Fig. 8.1 First building blocks in total quality

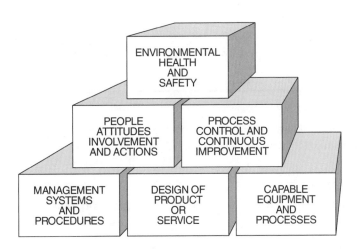

Fig. 8.2 Building blocks for a practical application of total quality

QA in Design

It is often forgotten that

> . . . the quality of the product or service starts and is intrinsically built-in (or excluded) at the very first design stage.

If the original design concept of the product or service is wrong or an element of the design is miscalculated, a wrong material or software selected, or the service conditions not fully defined, etc., the product/service will fail. No amount of quality assurance, process control or people involvement will prevent a plane falling from the sky if a strut collapses, or make a meal taste delicious if the original menu is incorrect.

Part of the QA Manager's role is to ensure that where the organisation designs the product or service that there are effective design procedures and that the design is also subject to formal reviews at appropriate stages. This review process should have input from the customer representative (or internal sales/marketing), purchasing, manufacturing and QA, to ensure:

- the design target (or service brief) reflects accurately the client's needs or specified requirements;
- the parts or materials specified are readily available at appropriate costs;
- it can be manufactured consistently to specification (over many batches or product/service cycles);
- that it can be inspected, tested and verified;
- the final product or service delivered at the end of the development process fully meets the customer's (or potential purchaser's) specified needs.

> The above requirements are obvious but it is surprising how many QA Managers do not get involved in, or are totally excluded from, the design stage.

Management systems and procedures

It is the QA Manager's most obvious role to ensure that quality is built into the product/service by effective management systems, procedures and controls at each stage of the 'process' (see Fig. 3.6, Chapter 3).

It is the author's opinion that the most effective method of achieving initial quality improvement is by implementing an effective management system based on ISO 9001:2000. However, I stress again that:

> Obtaining the registration certificate is secondary. The most important aspect is to get an effective, efficient and economical working management system for your particular organisation.

Because the above is not always stressed or the designated QA Manager is inexperienced or has not sought, or been given, sound advice,

> there are many systems which have achieved registration but are completely over-documented, bureaucratic, unfriendly and very expensive to run.

These systems are the ones always quoted by those making excuses for not introducing QA or not seeking registration.

It is the QA Manager's specific task to lead the firm through a programme to registration and then to continually strive for improvement thereafter. The QA Manager may also be able to include finance, personnel and sales activities within these management systems, (whilst excluding them from the registration scope and assessment, where appropriate).

The QA Manager's other specific role may also be to ensure that **product certification** such as the BSI Kitemark® or meeting regulations such as labelling for consumer safety, records to support self-certification (CE mark), food or hygiene requirements, etc. are met in full.

Capable equipment and process control

If some of the certification bodies' advertising material is to be believed, you have only to get your documented management system in control and registered to ISO 9000 to guarantee a good product or service.

Unfortunately it is just not that simple. In addition to your 'design' being correct, your processing equipment and raw materials must first of all be 'capable' of consistently producing to specification and then also the process must be kept under control.

The controls can range from a precision engineering machining workshop understanding and using SPC charts, to a restaurant controlling its suppliers of potatoes, fat-cooking temperatures, etc., to the data provided to and the training of staff on an information desk.

It is the QA Manager's role to ensure that capable equipment is provided and that appropriate process controls exist and are continuously improved.

People

The importance of trained staff at all levels and their effect on the quality of the product cannot be overemphasised.

> You may enjoy a meal in an exclusive restaurant with beautiful decor and delicious food. However, if the cutlery is dirty and the toilets are filthy you are unlikely to go there again.

Few restaurant owners appreciate that the success or failure of their business may be in the hands of the least appreciated and unseen members of their staff.

Conversely, I have seen a successful 3-year Total Quality (TQ) programme ruined by the production director overruling a decision of the quality manager by sending a customer a known defective product in order to meet a delivery before the end of the month to achieve an internally set financial target. The shopfloor in particular noted what they saw rather than what they were told. They perceived that senior management were not committed to quality at all.

However, I would stress one feature with respect to staff and Total Quality programmes. You will note in Fig. 8.1 that I have shown 'Design' and 'Systems and Procedures' on the bottom row. The practical fact is that you can work on either one of these two in **total isolation** and you can achieve a lasting improvement in the product or service.

> To work on the 'people' aspect on its own when the 'design' or 'systems and procedures' have not already been effectively addressed is a recipe for disaster, the building blocks will collapse.

Inexperienced practitioners are often persuaded to buy into quick-fix projects or programmes involving only people under the guise of 'Total Quality', 'Customer Care', 'Internal Suppliers', 'Investors in People', etc.

It does not take a lot of imagination to realise that you can make people enthusiastic for a short period of time using one of the proven team-building tools. However, in a few months they will get sick of singing the 'company quality song', if the goods are being returned from the customer because the design is fundamentally flawed or the management system leads to confusion or departmental conflicts.

> It is distressing that many government funded organisations in the UK are promoting 'Investor in People' as an alternative to ISO 9000 registration instead of an excellent and natural extension.

Environmental, Health & Safety topics

It is being found that these topics (see Fig. 8.2) are becoming more important and it is often convenient to build them into an overall Quality System. In fact in some industries it becomes essential because, although Health & Safety, etc. are not in themselves requirements of ISO 9000, some items such as COSHH, Electricity at Work Regulations, etc. are in the client's contracts and therefore must be introduced into the QA system to meet the requirements of contract review.

Again in exactly the same way as with the 'people' aspect, it will be *impossible to effectively sustain an environmental policy* without effective controls existing for 'design', 'systems and procedures', 'people' and 'process control'.

I can't perceive that one could get approved to ISO 14001 (Specification for Environmental Management Systems) without firstly implementing the fundamental management controls inherent within ISO 9000. In practice if ISO 9000 has been done effectively it may be relatively straightforward to build on the additional requirements of ISO 14001.

> However, you will not sustain an environmental system or Total Quality programme, without the other five building blocks in place. Without them the programme will collapse and fail.

Practical implications of the role in achieving TQM

I have described above the role of the QA Manager. However, the inexperienced practitioner will not realise unfortunately that the culture of firms varies considerably with the type of industry background, custom and practice, location, Trade Union attitudes, etc. Also they can change dramatically overnight with the departure or arrival of a very senior executive of either strong conviction or alternatively one of apathy. Also a take-over by another organisation can have dramatic change for good or bad. Some of the variations experienced by the author are as follows.

Holistic

Where quality of product or service is paramount. Despite all the 'hype' and publicity such organisations are very rare. If you are the QA Manager your task is to carefully and sensibly progress systematically in turn through the three stages shown at Fig. 8.1 to Fig. 8.2 to the European Quality Award Total Quality model of the type shown in Fig. 8.3.

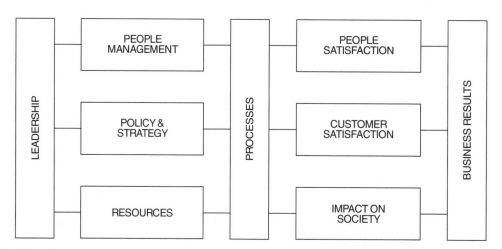

Fig. 8.3 The European quality award model

Do not make the classic mistake of going straight for the highest accolade shown in Fig. 8.3. You will fail!

Quality facade

Unfortunately this is very common with a company giving a facade of quality with

- high-grade brochures;
- outlandish quality policy statements (sometimes almost saving the world);

- beautiful foyers to their factories with perfect products on display, photos of sister plants in exotic places;
- appropriate directors or senior managers on Quality Committees or governors of Quality Centres, BSI Standards Committees.

The actual reality, if you can see their shopfloor, talk to their employees, talk to their customers or suppliers, audit their systems is often in stark contrast.

If you are unfortunate enough to find yourself as the QA Manager in such an organisation your role is clearly to strive to change the culture.

Production/output at all costs

Again despite all the propaganda, publicity, and information available on total quality, such firms are very common especially in the manufacturing environment where the processing activity is effectively totally hidden from the customer.

You usually find that these are run by a production manager/director who is a natural 'fire-fighter', 'instant fixer' or a 'hands-on manager'. Such individuals do not really want organised management systems or smoothly planned processes, because they thrive on strife, confusion and panic. UK manufacturers are often to be found panic-stricken and lowering standards during the last week of each month, getting the last extra pounds sterling worth of production to meet budget for the current month. They seem blissfully unaware that they may have destroyed all the MRP, JIT or any other planning system. Quality standards tend to drop even lower during this last week.

1.	Mandatory Quality Procedures (approved at the highest level). Certified and checked regularly by external and independent Lead Assessors.
2.	A formal and correctly structured Concession Procedure (that takes the production director/manager out of the decision process).
3.	Effective Audits: (a) Inside (b) External
4.	Knowledge of QA principles and techniques
5.	Knowledge of your product/service in detail Become the acknowledged "expert".
6.	Effective labelling systems.
7.	Very thick skin
8.	Careful use of external consultants to independently expose problems.
9.	Ability to smile!
10.	Above all: persistence!
11.	The Consumer Protection Act and other legislation.

Fig. 8.4   Tools of the Quality Assurance Manager/Department

The tools available to the QA Managers in this situation are those shown in Fig. 8.4. If the QA Manager is skilled in using these tools he can bring the quality of product or service into control. He may eventually be able to change the culture of the organisation and strive towards Fig. 8.1 or even Fig. 8.2, but it can be a long and lonely battle.

Practical common sense limited by perceived cash/budget restraints and culture

Every QA Manager would like to achieve the idealistic aim shown by the model in Fig. 8.3. In the majority of cases this will not be possible due to the environment, internal politics, vision or culture of the organisation. The QA Manager must not get frustrated by these barriers but must strive to continually improve.

With skill, connivance, patience and most of all persistence one can, over the years, have the satisfaction of achieving significant improvements in the QA system and in the product or service an organisation offers.

Finally, being effective in your role

To be really effective as a QA Manager it is essential to become knowledgeable in your craft. Unfortunately there are few university courses or training schemes that will create effective QA Managers. In fact the most effective QA Managers (or QA Consultants) have usually previously held senior positions in other disciplines. However, you will not become expert by reading a couple of books on Total Quality or attending an appreciation course.

I would strongly recommend that those who wish to be taken seriously as QA practitioners consider taking the IQA membership courses, the City and Guilds 743 in QA, or the Open University course material on Quality Systems and Techniques. All these courses will then give you a strong foundation to build on.

I don't know who to credit with the quote but it's worth remembering that

> 'Quality is unique in that you don't have to be sick to get better.'

Chapter 9
Advice on consultants

The selection of Quality Assurance Consultants is particularly fraught with danger.

There are no reliable figures or measures on the ability and success of QA Consultants but my very subjective assessment, based on many years' experience, will perhaps give you cause for concern. QA Consultants are approximately divided:

40–50% are excellent, good, conscientious, honest and provide sensible advice, sensible systems at a fair price. It is unfortunate that only a few of these have seen the obvious need (or found the time) to get themselves ISO 9000 registered.

25–40% are useless or even dangerous, they come in several categories:

- The early retirement manager who claims 'he once did a bit of quality', now looking for a paying hobby, or the 'bag of wind' looking for companionship or something to keep them interested.
- The Quality Manager or Inspector who has put **one** system in and got it registered who does not really know the standard and how to interpret it for different industries, and only knows a particular industry-based quality system. Hence you finish up with systems obviously 'lifted' from a large engineering firm forced onto a small machine shop (or, even worse, into a firm without a tangible product and providing a service) that are totally inappropriate and will not work effectively, efficiently or economically.
- The academic who tries to interpret and apply a 'hidden meaning' behind ISO 9000.
- The ex-accountant or PR consultant whose own market is barren and tries to get onto the 'quality bandwagon' or perceived 'gravy train'. Usually they try and tell you 'You are not ready for ISO 9000 yet', or 'Firstly you need consultancy for Training Needs Analysis', or 'QA Awareness' or 'Total Quality'. Or you should do 'Investor in People' first: see Chapter 8, for all the reasons why you shouldn't do Investor in People first.

The remaining 5% are extortioners either intentionally or by implication by their actions.

> Their whole intention is to milk the consultancy contract for as many days as possible. Hence you find throughout UK industry many completely over-documented systems that give ISO 9000 a bad reputation.

I have even seen two small firms with over 80 procedures! These firms both employ approximately 35 staff, one being a plastic injection moulder and the other a manufacturer of metal lockers. In the plastic injection moulder firm the consultant (?) has sold himself into the firm for two days a week for a whole year!!

> For your guidance, in the vast majority of firms employing less than 50 staff you should not need any more than 10 to 14 operating procedures. .

A good experienced QA consultant can, over a six/eight month period, document your system completely, liaise with your staff to ensure they are their procedures, desk top publish (DTP) any forms, assist you to implement your system and provide the first round of internal audits and hold your hand through assessment in less than 20 days consultancy. If it is a very small firm it can be done in less than 15 days work. It should be appreciated that for each day on site there is a day or another days work for the consultant off-site for documenting your system.

The fee rates applicable in the UK for a good QA consultant range from £250 to £500 per day (2002 figures). The appropriate figure in other countries can easily be calculated by ratios of currencies, etc. For less than £200/day they are probably very poor sole practitioners with attendant risks. (The sums just don't work out any other way; most QA consultants are lucky to get 100 to 130 truly productive, actual fee earning, days in a year. Hence £250 per day yields less than £30,000/year gross. Sounds good until you start subtracting the cost of telephone, stationery, secretarial costs, car lease and petrol, private pension scheme, health insurance, hotel and cafe bills, etc., this will reduce the income to substantially less than £20,000/year.)

The higher fee of £400–500/day should not be paid to any sole practitioner. This is the figure applicable to somebody working for a large corporate consultancy who typically will have a 30% overhead to carry.

> Certainly do not pay above £500 per day for any QA consultants.

> I have heard of people paying £600 to £700 per day. I can assure you that URS can direct you to three good consultants at substantially lower rates than these.

If you do the sums above (which most astute Managing Directors will have worked out already in their sub-conscious) I'm informing you that achieving registration to ISO 9001:2000, using a good QA consultant will cost between £4,000 to £12,000 (2002 figures).

In some cases recognised consultants may acquire some grant aid and reduce this down to £2,000 to £6,000.

> Hence it can be seen that there are significant savings to be made in doing it yourself and this book will pay for itself many times over.

Please be assured that this book provides **all** the information your require to achieve registration. However, whilst it provides the information and the model, you have to provide the effort and the time. Do be aware that typically this will take an average of 4 hours per week of your own time (or a designated departmental manager, graduate or equivalent) over 6 to 8 months.

However, in spite of the cash savings and the satisfaction of doing it yourself, some of you might still choose to use a consultant to develop and implement your system. However your understanding from this book will ensure that you stay in control of the project.

> I believe the most productive use of QA consultants is to do a 'dummy' assessment, provide internal auditor training or for a maintenance contract to carry out your internal audits and possibly document control.

How to choose a consultant

When seeking or 'fishing' for a QA consultant, you desperately need to sort out the 'friendly dolphins' from the 'sharks' and the 'bag of wind', see Fig. 9.1. The trouble being that they all look the same on the surface!

Fig. 9.1   The dangers associated with 'fishing' for QA consultants

Useful questions to ask are given here.

1. *'Are you, as a QA consultant, registered with an accredited certification body to ISO 9001?'* Just consider, if you got into a car for a driving lesson with a driving instructor and you discovered he hadn't passed his basic driving test himself, would you continue? Would you be satisfied he knew how to drive in theory? There is no basic difference. Why hasn't the QA consultant passed the test himself and become ISO 9001 registered? Don't accept 'because we are small' or 'it's not necessary'. If it's necessary and makes good business sense for you, it also applies to the QA consultant.

2. *'Have you some professional qualifications as a Quality Practitioner?'* Accept Associate Member of the Institute of Quality Assurance as a minimum. (Do note that licentiate, LicIQA, requires minimal, if any, experience in quality assurance. Also, affiliate membership is open to anybody. Do be wary of those claiming membership of the Quality Federations, these are often open to anybody.)

3. *'Are you a Registered IRCA Principal Assessor, Lead Assessor, Registered Assessor or have you passed the Lead Assessor Course?'* Do be aware there is no such animal as a 'Qualified' Lead Assessor, as I've seen claimed several times.

4. *'Have you a track record of success? Name the companies for whom you have been a consultant who are registered?'* I'd be looking for a list of at least six or more immediately provided from memory. *'Please provide me with the name and telephone number of three in similar industries to myself or firms of a similar size.'* Do ring them and see what they tell you.

5. *'Please allow me to look for a minute at a typical QA Manual and set of Operating Procedures you have done recently for a client, one from somebody in a different but straightforward industry so they are not a competitor and there is no confidentiality problems.'* If they cannot or will not show you such a Quality Manual and set of procedures (which in any case they could easily sanitise by removing the firm's name) the implications are obvious. You need only to look at the QA Manual and Procedures for a couple of minutes at the most to see what you need to know. They should be simple to understand and be

brief. They should be similar to the models enclosed. If they are jargonised, if they have complicated page controls with signatures on every page with transmittal sheets, etc. say thank you, but no. There should be at least a two-tier system of QA Manual and Operating Procedures, they should not have been combined (see notes on Chapter 5 why this is necessary). Are they too voluminous? There should be certainly no more than, say, 14 Operating Procedures. Look at the enclosed example of a Quality Manual and Operating Procedures. This is what yours will look like at the end of the project. Are you happy and feeling comfortable with what you see?

6. *'Are you the actual person who is going to provide the consultancy?'* You'd be surprised how many firms send someone different to the person you first deal with. If you are going to use a QA consultant, it is essential you interview and have the CV of that consultant. The personal chemistry must be right and you also need to feel you have confidence in him/her.

7. *'Please provide your fee rate per day including all expenses.'* Given that the consultancy is to document, assist, implement and advise up to and through assessment and address any corrective actions, what is the guaranteed maximum number of days that you will be charged for? What will be provided if we fail to achieve registration because of an error in the documented system?

It is unlikely that you will get somebody who passes all the above questions. However, these are the questions to put to the QA consultant on the spot and sort the 'dolphins from the sharks'.

I dare say that this chapter will cause some ruffled feathers among the less able QA consultants. However, it's not written for their benefit, it is to provide you with the information you need to implement your own system and hopefully to prevent you being misled.

Organisations in America and overseas should be particularly wary. There are several UK 'consultants' selling themselves just because they are British and hence they know all about ISO 9000. Some of these, in fact, have a very poor reputation in the UK, or in some cases no experience at all!

Also some are selling consultancy with an unaccredited 'Registration Certificate'. This is not allowed under UK UKAS rules; it's obviously not sensible to prepare somebody for the assessment and then carry out the assessment. Are they going to fail a firm when they have used their own consultant? What is an unaccredited certificate worth?

Quality clubs; group schemes

An alternative to employing a consultant directly is to go to a 'course' run by local colleges, Enterprise Agencies, Trade Associations, Chamber of Commerce, etc.

These typically include a series of 7 or 8 sessions of $\frac{1}{2}$ a day, held at two-weekly, or monthly periods, with perhaps a couple of $\frac{1}{2}$ day one-to-one consulting days at your site included.

These can be remarkably successful (I have seen two courses run by separate consultants achieve an 80% success rate of registration for attendees) or very poor. They do depend on the ability of the consultant or course leader. I have seen the tendency of some organisations to have an experienced consultant run two or three courses and then they 'acquire' the material and attempt to run the course themselves.

Before you join, ask the names of the firms on the first few courses and how many are now registered. Ring the attendees of previous courses and ask their opinion.

Obviously this book can easily be used as the workbook or, will provide valuable additional material for the text and models of self-help groups.

Appendix A

Useful names, addresses and telephone numbers

Below are the details of United Registrar of Systems Ltd. A leading world-wide accredited certification body, that I am proud to be a part of. They are, efficient, effective, customer focused, friendly, economical and totally professional. To be fair and so you can obtain a comparision of service and fees, I also have listed below details of six leading UK certification bodies.

United Registrar of Systems Ltd (U.R.S.)
Has its Administrative and Technical Head
Office at:
United House Station Road
Cheddar
Somerset
UK BS27 3AH

Operational and Sales Offices:

URS – UK
Glenview House
Courthouse Street
Pontypridd
UK CF37 1JW

Tel (0)1443 491129
Fax (0)1443 491124
sb@ffynnon.freeserve.co.uk

URS – France
8 Rue Monceau
25160 Chaudron
France

Tel 0033 3 81697802
Fax 0033 686 175282
urs.france@wandoo.fr

URS – Greece
Sof. Vempo & Messinis 2
165 61 Glyfada
Greece

Tel 0030 1096 27 700
Fax 0030 1096 27 822
ursmed@otenet.gr

URS – Italy
Villa Spinola
Via Corridoni, 5
16145 Genoa

Tel 0039 010 3627002
Fax 0039 010 3198546
ursitalia@tiscalinet.it

URS – Russia & Commowealth of
Independent States
Sergei Brodsky
Arbat 4/17
121019 Moscow

Tel &
Fax 007 095 291 5528

URS – North America
PO Box 372
Grabill
Indiana 46741
USA Tel 001 219 627 2633
Fax 001 219 627 3698
ursind@aol.com

URS – North America
365 River Chase Drive
Orlando
Florida 32807
USA

Tel 001 407 658 2666
Fax 001 407 658 7558
andyroweurs@worldnet.att.net

URS – Bangladesh
15/A Road 8
Gulshan-1
Dhaka
Bangladesh

Tel 0151 228 1282
Fax 0151 220 0706
rjmursbd01@hotmail.com

URS Ukraine
Kate Marchenko
LQC, Room 59a
2 Lazarenka Street
L'viv
79026 Ukraine

Tel 0038 (332) 95-02-45

URS – Japan
32 – 5 Imaizumi
Toyama City
Japan

Tel 0081 76 422 0991
Fax 0081 76 422 0992
urs-japan@mic.co.jp

URS – India
203.60 Upper Storey
Sadar Bazar, Muzaffarnagar – 250 001
Uttar Pradesh
India

Tel 0091 131 402580
Fax 0091 131 409030
mks616@yahoo.com

URS – Thailand
125/40 Moo 2
Soi Phetkaseam 68
Phetkaseam Road
Bangkae, Bangkok 10160
Thailand

Tel 0066 2 951 0305
Fax 0066 2 951 0306
pwattana@ksc.th.com

URS – Taiwan
7F-9, No 20, Lane 609
Sec 5 Chung-Shin Road
San-Chung City
Taipei County
Taiwan, ROC

Tel 00886 222 78 9822
Fax 00886 222 78 9823
johnw3@ksts.seed.net.tw

URS – Pakistan
Office No: 101, 1st Floor
Plot No: 7/C
6th Zamzama Commercial Lane
D.H.A. Phase 5 – Karachi
Pakistan

Tel 0092 21 583 7719
Fax 0092 21 5831099
urs@cyber.net.pk

URS – Singapore
7500-A, Beach Road
#06-323, The Plaza
Singapore 199591

Tel 0065 6297 3630
Fax 0065 6297 9379
quality.qms@pacific.net.sg

URS – South Korea
Vianet Bldg 3FL
74-4 Youngdeungpo-8 Ga
Youngdeungpo-Gu
Seoul 150 – 038
South Korea

Tel 0082 2 2636 9003
Fax 0082 2 2636 9070
esca@urs.co.kr

URS – Malaysia
No 11 2nd Floor
Jalan 1/24D, Wangsa melawati
533300 Kuala Lumpur
Malaysia

Tel 03 4142 8806
Fax 03 4142 8806

URS – Indonesia
Graha Kencana 8th Floor
Jl. Raya Perjuangan No 88
Kebon Jeruk
Jakarta 11530
Indonesia

Tel 021 5366 0660
Fax 021 5366 0661

URS – Saudi Arabia
PO Box 51141
Jeddah 21463
Kingdom of Saudi Arabia

Tel 00966 2 668 1371
Fax 00966 2 668 1370
urska@hotmail.com

URS – Turkey
Inonu Cad Zitas Is Merkezi
D 2 Blok Kat: 3
Daire: 8 Kozyatagi
Erenkoy, Istanbul
Turkey

Tel 0090 216 343 4787
Fax 0090 216 445 7360
metrotest@metrotest.com.tr

URS – Iran
PO Box No 19395-3769
Tehran, Iran

Tel 0098 911 230 69 67
Fax 0098 21 267 949
bagheri@mavara.com

The following are leading UK accredited certification bodies.

BSI/Quality Assurance
389 Chiswick High Road
London
W4 4AL

Tel 020 8996 7720
Fax 020 8996 7540

National Quality Assurance Ltd. (NQA)
Warwick House
Houghton Hall Business Park
Dunstable
LU5 5ZX

Tel 01582 539000
Fax 01582 539090

Lloyds Register Quality Assurance Ltd. (LRQA)
Hiramford
Middlemarch Business Village
Siskin Drive
Coventry
CV3 4FJ

Tel 024 7688 2261
Fax 024 7663 9946

Bureau Veritas Quality International (BVQI)
Tower Bridge Court
224–226 Tower Bridge Road
London
SE1 2TX

Tel 0171 378 8113
Fax 0171 378 8014

Det Norske Veritas Quality Assurance Ltd. (DNVQA)
Palace House
3 Cathedral Street
London
SE1 9DE

Tel 0171 357 6080
Fax 0171 357 6048

SGS Yarsley International Certification Services Ltd.
Trowers Way
Redhill
Surrey
RH1 2JN

Tel 01737 768445
Fax 01737 761229

Other useful sources of information are:

International Register of Certified Auditors (IRCA)
c/o IQA
12 Grosvenor Cresent
London
SW1W 7EE

Tel 020 7245 6833
Fax 020 7245 6844

(The licensing and approval authority for Principal Auditors, Lead Auditors, Auditors and Provisional Auditors. They can immediately confirm, if requested, whether an individual is registered, some who claim they are, are not. NOTE, there is no such animal as a 'qualified lead assessor' sometimes claimed by consultants or trainers who have done a Lead Assessor Course but have not enough experience to have achieved registration.)

Institute of Quality Assurance
12 Grosvenor Cresent
London
SW1W 7EE

Tel 020 7245 6722
Fax 020 7245 6755

(The only recognised professional body for QA practitioners in the UK.)

National Quality Information Centre
c/o IQA
Address as above.

Tel 020 7245 6669

(A very much under-used source of help,
advice and information. It is free and
unbiased and not trying to sell you
anything.)

**United Kingdom Accreditation Service
(UKAS)**
21–47 High Street
Feltham
Middlesex
TW13 4UN

Tel 020 8917 8400
Fax 020 8917 8500

(The government body in the UK
responsible to the Secretary of State, for
auditing, authorising or the 'accreditation'
of certification bodies. Will supply lists of
accredited certification bodies and their
scopes of approval.)

**National Accreditation of Measurement and
Sampling (NAMAS)**
Queens Road, Teddington
Middlesex
TW11 0NA

Tel 020 8943 7140
Fax 020 8943 7134

(The body in the UK authorised to assess
and certify Test Houses and Laboratories,
now part of UKAS. **NOTE** the approval is
not to a general scope. It is very specific to
the tests actually approved and checked. It
is therefore ESSENTIAL to carefully check
test houses before sending instruments for
calibration, to ensure they have approval
for that particular type of equipment.)

Stationery Office
London

Tel 020 7242 6410 (Also 0870 600 5522)
Fax 020 7242 6412

(UK official source of standards and data)
online www.clicktso.com

ILI
Index House, St. Georges Lane, Ascot
Berks. SL5 7EU

Tel 01344 874343
Fax 01344 291194

(A supplier of standards, can supply a copy
of virtually any standard from any country
in the world. Also provide an 'updating'
service to advise you of changes to any
standards you hold.)

Standards by Internet
(**Note:** standards can often be purchased
and downloaded via the Internet at
reduced prices. E.g. qualitypress.asq.com
standards.com.au

The Federation of Small Businesses
Sir Frank Whittle Way
Blackpool Business Park
Blackpool, Lancashire
FY4 2FE

Tel 01253 33600
Fax 01253 348046

(The representative voice of small
businesses in the UK. Have very correctly
been very vocal about the stupid
over-documented systems put in by some
consultants and the over bureaucratic
approach of some certification bodies.)

Appendix B

Practical definitions and explanation of terms

1st party	The original equipment or component manufacturer.
2nd party	A subcontractor or supplier under a direct contract to the OEM, or main contractor, both for manufacture or services, and for installation, repair and maintenance.
3rd Party	A party completely independent of the 1st and 2nd Party, e.g. an independent 3rd Party Certification Body (e.g. URS, BSI, NQA, LRQA).
4th party	Term, sometimes used to describe an independent but approved agent, e.g. a company who is authorised by a computer manufacturer to repair computers returned directly from retailers under guarantee and then invoice the computer manufacturer directly, is called an approved 4th party repairer.
Acceptable quality level	See AQL.
Advice note	See Packing Note.
ANSI	American National Standards Institute.
API	American Petroleum Institute.
AQAP	Allied Quality Assurance Publication (NATO).
AQL	Acceptable quality level. This is very confusing as it is actually a measure of the number of **defects** that a production process or delivered batch of goods is allowed (see also sampling). Usually expressed as a percentage, e.g.,

$$0.4\% \text{ AQL} = 0.4 \text{ defects per } 100$$
$$= 4 \text{ defects per } 1,000$$

ASQC	American Society of Quality Control.
Attributes	The characteristics of an item that either exist or are missing, e.g., two ears, four wheels, 24 keys, clear surface without scratches, no foreign bodies, no dents, better than approved visual standard, passes through gauge. See 'Inspection by Attributes'.

Average

A measure of the central tendency of a population or sample (see also mode and median) calculated by adding the values of all of the samples and dividing by the number of samples observed, e.g. average of $9 + 3 + 4 + 5 = 21 \div 4 = 5.25$.

Be careful of tests or products claiming 'good average results'. Bags of sugar of average weight of 1 kilogram may be made to 0.99 to 1.01 kilogram or 0.8 to 1.2 kilogram. If you had the individual bag of sugar weighing 0.8 kilogram, it would not impress you at all that the 'average' weight is 1 kilogram. An average should always be stated with a measure of dispersion (see Range and also Standard Deviation).

Bath tub curve

Well designed mechanical, electrical and electronic products tend to follow a very predictable pattern of probability of failure throughout their working life.

If plotted as 'probability of failure' v 'life' (with life on the bottom, horizontal axis) the shape of the curve is that of a 'bath tub', with 3 distinct phases:

(i) Infant mortality – early breakdowns due to errors in design, manufacture or wrong application, with probability of failure decreasing as life in service progresses.

Responsible electronic manufacturers often try and remove these failures by an initial overload 'burn-in' during manufacture.

(ii) A more or less consistent and very low probability of failure – resulting from random problems or accidents.

(iii) An upward period of wear out – as items reach their design life, and failures increase more rapidly.

Note: Just think, next time you are in your electrical retailer/ superstore, why is the salesman spending so much effort and persuasion to get you to take an extended warranty, is he covering you for (ii) or (iii)? Your manufacturer's guarantee should already cover you for (i).

Bell curve

See Normal Distribution.

'Bought-in-and-supplied on'

See Chapter 6.

Bonded store

A secure place in which items that have passed test or have unique features are held, that need formal authority to release. **Not** a requirement of ISO 9001, only introduce if needed for your processes.

Break-even curve

Often the Quality Practitioner is trying to persuade people to spend more money on items to 'improve' the product. It is as well that the Quality Practitioner is aware of the hard facts of life.

The 'improvements' have to be paid for and, therefore, enough units sold to recover all the costs.

The improvements will not pay for themself until:

(a) enough units have been sold at X price per unit for the total sales income to be higher than. . .

(b) initial set-up costs + fixed overhead or constant costs + variable costs per unit.

(a) and (b) can be produced as two simple straight line graphs and where (a) [sales income] goes above the sum of (b) you start to make a profit. This is called the break-even point.

Breakdown inspection	A very powerful verification tool for checking the process or design. A sample is taken at the end of production and critically tested and broken down to check effectiveness of processes and controls. Similarly samples are purchased back after 'x' months in service, tested and broken down.
BSI	British Standards Institution.
Burn-in	See Bath Tub Curve.
Calibration	Common term to describe ensuring all test and inspection equipment is correct or standardised, usually by checking against approved standard checking pieces that are traceable to national standards, where these are available. In appropriate cases equipment may be simply checked to the manufacturer's instructions.

Also a generic term used to describe the activities contained within Clause 7.6 of ISO 9001.

Cause and effect diagram	Also known as Fishbone or Ishikawi diagram.

A very simple but very powerful technique to use. It can prevent people erroneously taking action on the 'effect' of a problem rather than finding the root cause(s).

Basically the 'effect' is written on the left hand side of a page and a 'fishbone' constructed on the right of **all** the possible items that would cause this effect. These causes are then considered and eliminated in turn.

A technique that must be learnt by any self-respecting QA Practitioner or Manager.

CE mark	This is a prominent mark on the goods showing clear visual evidence that the items comply with essential or minimum legal and/or safety requirements of the appropriate European Community Directives. Can be misleading as the mark implies only compliance with minimum legal safety requirements, e.g. nothing to do with quality or grade. Also some items have to be independently tested and certified by a Notified Body. With others, it is a case of self declaration and records to be held by the manufacturer or importer.
CEN	European Committee for Standardization (Comité Européan Normalisation)

Certificate of batch (or lot) traceability	Items manufactured, assembled or tested in discrete numerical lots (or date) under uniform conditions and remaining identified. Mainly used to permit future product withdrawals.
Certificate of conformity	A document signed by a representative of the company confirming that the product or service meets the stated requirement. (These used to be a popular demand or request of purchasers. Now realised by most people that it gives you no more than your entitlement under your agreed contract, i.e. an additional piece of paper of no particular value. Also produced by your supplier by simply pressing another key on the computer. Certainly not worth paying an additional fee to obtain!)
Certificate of material traceability	A certificate tracing the history of the material back to a Certificate of Material Analysis from the manufacturer, foundry, forge or chemical works usually controlled with a batch, lot or melt number (see also Certificate of Material Analysis).
Certificate of material analysis	A document signed by a qualified person identifying the parts or constituents of a substance or material sample. (It is essential that the material analysis is from the material or 'pour' or ladle of your cast items (or forge bar for forgings, pressings), especially as some of the additives in the melt deteriorates through the day. A 'day of melt' or 'typical' analysis sometimes provided are of little value. If you have a critical piece and you wish to be sure of the constituents you may wish to design the item with an additional lug that you can saw off after delivery and have analysed at a laboratory of your choice.)
Certificate of test	A document confirming that an item or material has been tested and passed the specified test, or gave the results listed. To be of value needs to be signed by the individual who actually carried out the test. Of more value if this individual is totally independent of the 'production process'. If complete independence is required URS or most certification bodies will provide a suitably qualified inspector to witness tests.
Chilled design	A design that has completed all its design and development phases and is correct in principle, but is not yet finally settled on all the details or tolerances. Possibly awaiting field trials, etc.
	Some of the design drawings and specification may be given limited approval to issue as a 'chilled design'. It may be necessary to do this in order that materials with a long delivery time can be ordered, or to enable others to design service facilities to be ready for the in-service date of your equipment. See Fig in Appendix to the draft Quality Manual. A very useful device to issue provisional drawings whilst keeping the project under careful and documented control. (See also frozen design.)
Concession	Written authorisation to use or release a quantity of material, components or stores already produced which do not conform to the specified requirements.
	Also sometimes described as a 'waiver', 'deviation' or 'variation'.

In less formal systems (typically for small firms) this will also include 'production permits'.

Cusum graph

A simple horizontal graph of a plot of the observed value minus the planned (or reference) value continuously added together (e.g. the cumulative sum). If the mean does not change the line stays approximately horizontal with the results scattered each side of the line. If it starts to go progressively upwards or downwards it shows very clearly the mean is drifting.

Contract

In the UK an unconditional acceptance of an offer, or tender, to provide goods or services, that are not illegal, for a consideration (e.g. money or goods).

The idea that somebody places an order on you is misleading. Thus a thick 'contract document' is, in fact, a complicated offer, or invitation to tender. Also you cannot accept a contract subject to conditions. You are, in fact, making a counter offer.

You should note that the one who unconditionally accepts the offer, accepts the terms and conditions of the other party.

Also ensure there is always a consideration. Do not accept goods or services or part of them (e.g. service manuals or installation) free of charge. Always insist on paying at least £1, otherwise there is no contract and hence no come back. For example if somebody offers to sell you a complicated piece of equipment for say £18,000 with 1st years service and the maintenance instructions free of charge, I recommend you actually order equipment at £17,800, 1st year service at £100 and maintenance instructions at £100.

Control charts

Charts of results taken at regular intervals to check or control a process.

The most common being a combination of two charts '$\bar{x}$' and 'R' That is:

$\bar{x}$ = The mean, or average, is checked to see if the mean, and hence the process, is drifting out of tolerance, e.g., caused by tool wear;

and

R = Range to ensure that the outer limits of capability (say + or − 3 Standard Deviations) of a process are not spreading wider and hence allowing out of tolerance parts to be produced, e.g. caused by loose bearings or tailstock, hardness of material changing.

Controlled document

One that is required to ensure that the quality system and customer's contract requirements are met. Must have the following:

(a) identification
(b) issue status or date of issue (may have both)
(c) a signature authorising the issue

Items (a) and (b) must appear on each page. Item (c) may appear on the document, however, it is usually more convenient to have the authorisation on a single control sheet or record book.

Corrective action

Short term or immediate 'fire fighting' action to correct an error, non-compliance, customer complaint, and prevent re-occurrence (see also Preventive Action)

Cost of poor quality

See Chapter Two.

Customer-supplied product

see 'Free Issue'.

Design control

See Appendix F. Also Design Input, Design Verification and Design Validation.

Design input

A documented, agreed and approved statement of what performance or output the design should eventually arrive at. Sometime broken into two distinct stages, e.g. 'Design Target Specification' for initial design concepts and 'Design Performance Specification' for the final design requirement.

Any designer who starts producing a design without such an agreed/approved design input specification is very unwise. In such a case, if the finished design does not perform, it will be automatically his/her fault, regardless of whether the market ing concept, assumptions, identified unknowns/uncertainties or figures provided were completely wrong.

Design review

A formal, comprehensive and recorded critical review and criticism of a design by representatives of the appropriate bodies, e.g. the user, the customer or main contractor, sales/ marketing, manufacturing, purchasing, quality, etc., to check the proposed design is:

(a) what is needed
(b) whether it can be made
(c) what can be inspected
(d) is the raw material available, etc.

Note: Design Review is a mandatory requirement of ISO 9001: 2000.

At each Design Review do ask each time the most important question of all ' is this project still viable?' (I have never yet seen a project manager volunteer to kill his own project).

Design verification

A check on the 'design' to ensure the 'design output' meets the specified 'design input', e.g. drawings and calculations checked by others. (See also Design Validation).

Design validation

Trials and tests defined as representing actual service, consumer or field use, carried out on the final product (or service).

May also be carried out on sub-assemblies or elements of a design.

(See also Design Verification).

Deviation	See 'Concession' or 'Standard Deviation'. (Two completely different meanings.)
Discrepancy	See 'Non-compliance'.
DIN	Deutsches Institut für Normung (German Standards Institute).
Disposition	The act of disposing of an item, e.g. scrapping, shredding, incinerating. A requirement to address in ISO 9001, as far as 'records' is concerned, that is often erroneously omitted.
DQS	Common abbreviation for 'Documented Quality System'.
DTI	Department of Trade and Industry.
Eighty/twenty rule	See Pareto.
EMAS	Environmental Management Systems, to achieve environmental protection by recycling, purchasing 'green' products and conserving energy. Various environmental schemes and standards are now becoming relevant, the most widely recognised being ISO 14000.
EMC	Electromagnetic Compatibility (designing and testing of electrical/electronic equipment to ensure that operating one does not affect the other, e.g. vacuum cleaner not affecting a TV picture, microwave oven not affecting users's Heart Pacer, etc.).
EN	European Normalisation (see CEN).
EOQC	European Organisation for Quality.
Evaluation of subcontractors	That you can show (to meet the requirements of Clause 7.4 of ISO 9001) that your subcontractors (e.g. the organisation or people from whom you buy goods or services) have been meaningfully evaluated to show why you approve of them as one of your subcontractors. Typical methods:

(a) Can demonstrate that this supplier is ISO 9000 registered with a UKAS approved body for the **scope of the product you intend to purchase**. (It is most important that you check this scope very carefully. One of the largest UK certification bodies issues an impressive presentation certification emphasising their logo but completely omitting the approved firm's scope, this being detailed in an insignificant looking annex.)

(b) There is historic and documented evidence of consistently supplying the relevant services or goods to specification. If you claim this, do expect the assessor to check that this is the case.

(c) That there has been a documented and recorded evaluation of the supplier to confirm he can supply the goods/services (CONSISTENTLY if a long term trading relationship).

Typical evaluation could be: (i) trial purchases carefully inspected before use, (ii) audit visit with a favourable documented report, (iii) documented recommendations by others.

Factored product

See 'Bought in and Supplied on' (Chapter 6).

Fishbone diagrams

See Cause and Effect diagram.

FMEA

Failure Mode and Effects Analysis. A powerful, but simple technique, to look at each element or component in a design and see the different ways it can fail. You then assess the probability of it failing and whether it is an acceptable life in service (or safe mode of failure) or if it can, or should, be re-designed.

Free issue material or parts

Also known as 'Customer Supplied Property' (was also confusingly known as 'Purchaser Supplied Product' in the 1987 version of the standard).

Material or parts that are supplied 'free of charge' by your customer for incorporation into their product or service, or are necessary for you to make, test or specify the product.

Items that are sent by a client for you to repair will also, by definition, be 'free issue'.

Note: Often erroneously 'written out' of DQSs as not applicable to the activities of . . . . plc. Do be careful as a lot of firms, without realising it, do have 'free issue' materials, services, drawings, specifications, tools, equipment, etc. (See Chapter 5.)

Frozen design

A design that has completed all its design, development, field and production trials. A design that is ready to be approved and issued unreservedly into manufacture or service. (See also 'Chilled Design'.)

Gage

USA spelling as against UK spelling of 'gauge'.

Gaussian curve

See Normal Distribution.

'Go' Gauge

A gauge precision made to the highest size (normally a diameter) of a tolerance band. All items to be acceptable must 'go through' the gauge or they are oversize. For checking holes this is usually a plug gauge made to the lowest permitted internal tolerance. The plug gauge must 'go through' the hole or the hole is too small.

Grade

An indicator (possibly numeric), or category relating to strength, status, class or cost. Not necessarily related to quality (see Chapter Three for definition of Quality).

Hardness testing

Test usually carried out by pressing a small hard ball or pyramid into material under a known calibrated load. The width or diameter of the impression is measured via a microscope and, by reference to tables, the hardness of the material is given. As tensile strength and other features of materials are often related to hardness it can provide a quick first check on material characteristics, say at goods inwards inspection.

Hawthorne effect	Named after an experiment conducted at Western Electric Company, Hawthorne Plant in Chicago in the 1930s. Basically improvements were made in workers' conditions, heat, light, rest periods, length of day, etc. It was found that output efficiency and reject rates improved as the environment improved. However, they continued to improve when the environmental conditions were reversed. These were attributed to the improved industrial relations and the fact that management were actually showing an interest in the wellbeing of their shopfloor staff. (Having spent 6 years on the shopfloor before going to University, I find this a rather obvious observation but it is still widely quoted as being very profound.)
Histogram	A chart giving a frequency distribution. Made of bars of equal width. Often very worthwhile doing on a meaningless set of data. The distribution picture often reveals vital information, e.g. incapable processes, skewed distributions, a sifted sample at goods inward inspection, sample off two different machines, etc.
Hold	See Quarantine.
IAF	Accreditation agencies of repute and international standing, such as UKAS, have signed-up to the International Accreditation Forum (IAF), to ensure that the rules governing Registrars and Regulations by the accreditation agencies are consistent throughout the world.
	Consequently, in 1998, a multinational agreement (MRA) was signed by seventeen accreditation agencies around the world, including UKAS, whereby each of the seventeen agencies confirm mutual recognition of each other's accreditiation. As a result, any certified client by a Registrar accredited by one of the seventeen Accreditation agencies that have signed-up to the MRA, can be assured that their registration will be accepted in any country belonging to the MRA.
Identification	A serial number or mark that uniquely identifies an item, lot or batch.
Independent third-party certification body	URS, NQA, BSI, Bureau Veritas, Lloyds Register, etc., who carry out assessments of firms' quality systems and award certificates confirming approval to ISO 9000. (Beware unaccredited bodies, or bodies claiming accreditation not recognised by the IAF.)
Inspection	Activities such as checking, measuring, gauging, examining or testing one or more characteristics of a product or service and comparing with a standard or the specified requirements.
Inspection and test status	A positive indicator (by mark, storage location, label, card, etc.) that items are:

(i)   awaiting inspection
(ii) quarantined
(iii) rejected
(iv) passed inspection or test, etc.

Inspection by attributes	Inspection where characteristics of an item are checked to see if features exist or are missing to assess whether they are either conforming or not conforming to the requirements of the product or service. (See also Sampling.)
Inspection by variables	Inspection where characteristics of an item are evaluated against a numerical comparison, measurement or scale. Sampling plans based on variables are quite complicated. These are often simplified to inspection by attributes by putting the measurement into acceptable tolerance bands and asking if the items have the attribute of laying in that tolerance band or not. Often further simplified by using 'Go' or 'No go' gauges.
'Investor in People'	An award introduced by UK government to encourage organisations to invest in individuals' training. Has had some success as initially given much funding through local agencies. Undoubtedly many individuals have benifited. However, the award is based on the assessor's subjective opinions and visits are only every 3 years.
IRCA	International Registrar of Certified Auditors. The UK register of Approved Lead Assessors and Assessors of QA Systems. (See Appendix A)
Ishikawi diagram	See Cause and Effect diagram.
Kitemark	A mark (Note: the property of BSI) that is permanently put onto a product to signify that the product has passed type tests thereby ensuring the product design and samples of the finished product conform with the relevant British Standard. Other organisations have product certification marks but the Kitemark is by far the most well known and recognised. (Note the CE Mark does NOT ensure such product conformity.)
Management review	A formal report to, and an evaluation by, top management of the status and adequacy of the quality system and the results arising. Note: A mandatory requirement of ISO 9001.
Mean	See Average.
Measure	Direct comparison to a physical standard (e.g. National Physics Laboratory standard) (see also monitor).
Median	A measure of central tendency (see Average or Mode). The middle value when the values are arranged according to size. Sometimes used for subjective examinations such as colour, taste or smell.
Mode	A measure of central tendency (see Average and also Median). The mode is the value that occurs most often, (e.g. 'a la mode' the most popular or fashionable) shirt collar size 16, shoes size 7, 25 mm pipe fitting, etc.
Monitor	An unspecified check, e.g. an observation or judgment (see also measure).

MTBF   Mean Time Between Failures.

MTTF   Mean Time to Failure.

Murphy's law   A humorous but pessimistic view and philosophy of life. Whatever can go wrong will! E.g. toast will always drop butter-side down, the delivery will be late, it will rain on the wet paintwork.

NACCB   National Accreditation Council for Certification Bodies (title now changed – see UKAS).

NAMAS   National Measurement Accreditation Service (now part of UKAS). The body responsible to the UK government for assessing, surveillance and approval (e.g. accreditation) of laboratories who provide a service of calibration of instruments and equipments, traceable to national standards. **Note**: Laboratories do **not** get a general approval of their Quality System (e.g. as per ISO 9000 certification) by NAMAS. They are only approved for very specific tests on specific materials or equipments. It is important to realise the distinction when using a NAMAS-approved laboratory, to ensure they are approved for your particular test.

NDT   Non Destructive Testing. Tests done on materials or components without impairing their future use. Can be either tests to check for flaws or defects or, to ascertain physical properties, e.g. hardness, conductivity, elastic constants, etc. Tends to require expert or, at least, specialist training. Typical techniques include, X and gamma rays, eddy current, dye penetrant, pulse echo, wall or lining thickness, etc. **Note**: If using an outside specialist, certification to ISO 9000 or approved by NAMAS is certainly appropriate.

'No Go' gauge   A gauge precision-made to the lowest size of a tolerance band. All items to be acceptable must 'not go through' this gauge or they are under size. For checking external round or cylindrical items a No Go gauge shall be a 'gap' or 'C' gauge (rather than a ring gauge), in case the item is oval and on size at 0°–180° and undersize at 90°–270°. The item being presented to the gauge at 0° and 90° to confirm it is No Go. Similarly a No Go plug gauge for checking holes (made to the largest size of the tolerance band) should have 70%–80% of the material removed on either side to make the gauge a slot or pin. This again should be presented at 0° and 90° to make sure the hole is not oversize, by being oval.

Non-compliance (n/c)   The non-fulfilment of a specified requirement of a quality system, quality procedures, instructions or contract (sometimes called a discrepancy). During formal assessments by certification bodies these are usually divided into minor or majors. A minor n/c is an isolated lapse, e.g. one work instruction not signed, one gauge not calibrated, etc. A major n/c is where there is a total failure of one part of the system, e.g. a required procedure missing or not implemented.

Non-conformance   The non-fulfilment of the specified requirement of a product or service.

Normal distribution	Also known as 'bell-curve', 'bobbys'-helmet curve' or 'Gaussian curve'. If we try and machine an item to 25 mm on a lathe they will not be all exactly to size. Some will normally be slightly smaller or slightly larger. A few will be even smaller and even larger. A very few will deviate even further away from the target 25 mm, either bigger or smaller. If these results were ranked into discrete values/intervals (of say 0.1 mm) and plotted as a histogram or tally chart of the occurrence of each size, the graph produced will resemble a bell shape with the majority of the results gathered around the 25 mm. The resulting shape of the curve is called a Normal Distribution, for the very reason it is 'Normal'. This shape occurs repeatedly in nature, e.g. height of males in a geographic population, neck sizes, shoe sizes, weight of fully adult male newts, etc. The only thing that varies is the width of the curve or spread of the results. This shape of the curve has been described mathematically and hence the spread (or measure of dispersion of the results) can also be described quite precisely and is called the Standard Deviation. Having taken enough samples of a particular population it is possible to calculate the Standard Deviation and hence calculate and predict very accurately the percentage of the population that will be within a particular size, weight or dimension range. The same rules apply for the majority of manufacturing processes and hence it is possible to predict the results (and defect rates) of processes by using these Statistical Process Techniques. Texts that make these principles simple to understand are *Right First Time* by Frank Price (ISBN 0-7045-0522-3) and *A Practical Approach to Quality Control* by Rowland Caplin (ISBN 0-09-147451-5). (See also Standard Deviation and SPC.)
OEMs	Common abbreviation for 'Original Equipment Manufacturer'.
Operating procedures	Written procedures on how one runs the business. See Fig. 5.1 in Chapter 5 to see where placed in the hierarchy of documents.
Packing note	The paperwork from the manufacturer or supplier of the goods delivered. Usually attached to, or packed within, the box, bag, envelope or container. Should detail the description and quantities of the items within the packs. Also called an Advice Note. **Note**: A very common mistake is to carry out goods inward inspection against the supplier's packing note only. This is a very grave error. Always inspect against the latest amended copy of your own purchase order, to ensure what they have delivered is, in fact, exactly what you ordered. Some stockists (typically, for example, stationery, food, engineering fasteners, anti-friction bearing, or other consumables) will send what they judge to be 'nearist equivalents' because that is what they happen to have on their shelves.
Pareto analysis	Said to arise from an Italian economist who observed that 80% of the wealth of the country was in the control of only 20% of the population. Similarly it is often found that 80% of all problems, rejects, defects, etc. are attributed to only 20% of the possible causes. Also 80% of sales are likely to be generated

by only 20% of the product range. Typically an exercise is done to identify the possible causes of rejects and then produced as a simple histogram or tally chart of the number of occurrences against each possible cause. The histogram/tally chart is rearranged with the largest to the left and then in order of decreasing values. This clearly shows which cause to tackle first in order to make an impact, e.g. to sort the 'important few' from the 'trivial many'. It is often found that the charts clearly show that 80% of the occurrences are very much to the left of the chart. Often called the 80/20 rule. Also applied to cost reduction exercises regarding losses and their causes. A very simple but very powerful and effective rule, and essential for any QA practitioner or manager.

Preventive action	Long-term preventive action to prevent a non-compliance occurring in the first place. (See also Corrective Action.)
Process	In terms of the ISO 9000 standard it simply refers to the process of taking raw materials and working on them and producing a usable output, e.g.:

Metal in → 'process' → car engine out.

Information and specified requirement in → 'process' → design of a building out.

Order to waitress in → 'process' → meal on table out.

Process capability	This is the ability of a process to reproduce consistently within a known, measured or calculated capability, limits or deviation. (See Standard Deviation and Normal Distribution.) It is generally acceptable for a process to be 'capable', the required tolerances must not be less than ± 3 standard deviations. Some firms are now striving for ± 6 standard deviations.
Product certification	See Kitemark and CE Mark.
Product liability	The responsibility on a manufacturer, importer, supplier, designer, licensor of technology or others to make restitution for loss or harm related to personal injury or property damage caused by a product or service.
Production permit	Formal authorisation, prior to production, or provision of the service, to depart from the specified requirements. In less formal systems can become classed as 'concession'.
Purchaser supplied product	See 'Free Issue'.
QAR	Common abbreviation for the quality assurance representative – the nominated or designated authorised representative of a company or organisation.
Quality	In general, the totality of features or characteristics of a product or service that bears on its ability to meet a stated or implied need. With regard to ISO 9000 it is easier to think of Quality = **exact** Conformance to drawing and/or specification. (See also Chapter Three.)

Quality assurance	Planned and systematic management actions throughout the organisation that are necessary to provide adequate confidence that the product or service will satisfy the specification.
Quality audit	An independent examination to determine whether the subject activities and the related results comply with planned arrangements and whether these plans or procedures are implemented effectively and achieve the specified objectives.
Quality circle	A Japanese technique of getting people to work together for quality, product, self or team improvement. Membership is supposed to be voluntary with members choosing the problems to work on with no financial reward. Became a vogue in the 1980s in the UK with very variable success rates. The management must be prepared to invest in training the teams in basic QA and problem-solving skills. Also UK employees do not appear to have the overwhelming desire for peer approval exhibited by Japanese workers. Neither, in general, do UK workers have the same overwhelming desire to truly and freely help or respect their employers (perhaps because it is not reciprocated). The author and many others, have found it better to use the same excellent basic problem-solving techniques and adapt to multi-disciplined project groups, working on priority or urgent problems set by management.
Quality control	Operational techniques at the shopfloor that are used to control manufacturing or service operations.
Quality costs	See Chapter Two.
Quality manual	A document setting out the general quality policies, of an organisation (also sometimes more correctly called The Quality Policy Manual). Often called the 'level one document', with the Operating Procedures being level two and any job-specific Work Instructions being level three. (See Fig 5.1 in Chapter 5). Should 'mirror' the requirement of ISO 9001 in simple language as it applies to the particular factory or organisation.
Quality plan	A formal controlled document setting out the specific process and appropriate inspection activities and resources, set out in sequential steps relevant to a particular product, service, contract or project. Can be produced as a document, a table or a flowchart.
Quality policy	A statement by the executive of an organization of the overall intentions and direction as regards quality. It should be signed, dated and a controlled document. Must be communicated to all staff (typically by placing on permanent display, issuing with pay packets, printed on the back of business cards or bottom of compliment slips, in small font on the bottom of every work instruction, etc.).
Quarantine	Goods or items held pending a decision on their use or disposition. Often applied to 'suspect' components or product. Also shown on some manufacturers cards as 'hold'.

Quarantine store	A segregated place to store items. These may be items that are awaiting proof that they comply with specification or alternatively that they may be known defective items awaiting disposal instructions.
QUENSH	Acronym for QUality, ENvironmental, Safety and Health. A rather poor acronym for a very sensible proposal. Having got your manufacturing and service processes well and truly under control and your DQS certified to ISO 9001, you are well advised to extend the same philosophy and controls to your health and safety requirements and also to the control of your environmental concerns.
Range	A measure of dispersion. Strictly speaking the lowest value subtracted from the highest value. However, with control charts, the 'range' may be a calculated figure. See also Standard Deviation.
Regulatory requirements	Any legal requirement established as necessary in a particular country, with regard to a particular product.
Reliability	The probability that items will perform correctly under defined conditions for a specified period of time at a given rate of failures.
Repair	Additional work to bring an item to a usable state but not necessarily to specification (as against Rework). Should only be permitted with a concession or other customer (or design department) approval (e.g. typically a weld repair on a shaft).
Resources	The items required to perform a particular task, contract, project or programme. Can be conveniently broken down into: • Personnel (including training and necessary experience) • Money • Equipment (including raw materials and also production and test equipment) • Technology (the design of the product or service, with the experience and knowledge to provide it) • Management system and controls (the lack of which dooms many a project) • Time (the resource most commonly denied or ignored)
Rework	Additional work to bring an item back to specification (as against repair).
Right first time	See Zero Defects.
Sampling	The technique of taking a representative sample from a population and deducing or estimating facts about the population. Typically acceptance batch sampling. To be useful they need to be based on sound statistical techniques based on the laws of probability. (See also AQL and also Sampling Plan.) The common practice of a sampling of 1 in 10 is

completely flawed. If you have a population of 10 and take a sample of one that is found to be defective, what does it tell you? Only that one out of a population of 10 is defective, the other nine may be good or bad. Also, it does not give you a decision whether to accept or reject the batch, e.g. what is the AQL? Also, if you have a population of 10,000, do you really need to take a sample as large as 1,000 to get a representative sample? To be meaningful use must be made of the appropriate standards, e.g. BS 6001, MIL-Std-105D, etc.

Sampling plan	Usually a table giving sample sizes and how many defects can be allowed in a sample to reject the batch i.e. the acceptance/rejection criteria. (See Sampling and also AQL.)
Service	Easily confused with 'servicing' with regards to ISO 9001; it is best to consider that these are not related at all. A 'service' in the context of ISO 9000 (other than Clause 4.19 of the 9001:1994 version) is a noun. For example, I provide you a service of producing drawings to your specification, or a service providing meals on wheels, etc.
	In order to interpret the standard correctly (as it was originally written for manufacturing or process industries) it can be useful to simply think of the 'service' you provide as a 'product'.
Servicing	'Servicing', as per Clause 4.19 of the 1994 version of the standard (corresponding to para 7.5.1(f) of the 9001:2000 version), is a verb and is the act of providing maintenance and repair to equipment that you have designed or manufactured. (**Note**: However, if your key or only activity is servicing equipment you have not designed or manufactured, this should be documented in your system to meet the requirements of Clause 7.5 Process Control, e.g. the process of providing a service of installation, service and repair.)
Shelf life	Items that must not be used beyond a certain date, i.e. they go out of storage 'shelf life'.
Sod's law	See Murphy's Law.
SPC	Statistical Process Control. The use of standard statistical techniques and charts to ensure items are reproducible on a particular equipment and then ensuring that such processes remain under control. By use of such charts it is possible to predict that items are going to be produced within or out of tolerance and to readjust the process before defects are made. (See also Standard Deviation and Normal Distribution.)
Special processes	Process that cannot be readily checked after completion to ensure that the specification has been met. Typical example would be welding, soldering, painting, application of liquid fertilizer, inspection of foundations prior to pouring of concrete.
Specification	A formal document that describes in adequate detail the requirements with which the product, material, service or process has to conform. This may include attributes and variables.

Squiggle list

Controlled lists showing appropriate staff's signatures or initials. (It can be embarrassing if an assessor asks who has signed or initialled a document and nobody knows who it is. I assure you it happens!)

Standard deviation

A measure of dispersion of the numbers or sizes of items arising in a population that follows a normal distribution. This can be calculated from a formula, but is usually read from tables or simply read off a calculator (or computer) after entering the values of the sample taken from a population. It is a fact of life (similar to the FACT that the ratio of circumference to diameter = $^{22}/_7$) that in a normal population 99.73% of the population will be within ±3 standard deviations.

It, therefore, follows that if the tolerances on a manufactured item are set lower than ±3 standard deviations the process is totally incapable of producing to tolerance. E.g. the normal distribution of a particular lathe when machining a metal at 25 mm diameter has an inherent +3 standard deviation and −3 standard deviation at + or −0.08 mm. That is, 99.73% of what it makes will normally be between 24.92 to 25.08mm. This is the natural or normal distribution and capability of this machine. It will always, unless there is an outside or special cause, follow this pattern. The machine has no knowledge of, nor can it guess, what is in the head of a designer. If the designer has, therefore, specified ±0.05 mm as required, (e.g. everything to be between 24.95 to 25.05 mm) the machine is totally incapable of producing consistently to this tolerance (i.e. it will always produce a predictable percentage of defects) regardless of the skill or dedication of the operator. As the majority of UK managers or supervisors do not know the capability of their machines, they will needlessly blame the operatives. (See also Normal Distribution and SPC.)

Standby redundancy

An additional (or redundant) piece of equipment that is provided in case of failure, e.g. spare wheel in luggage compartment (boot) of the car, auxiliary generators for critical production processes or hospitals.

Statistical techniques

The use of the theories of mathematical probability to estimate the defect rate in large batches of components or, the predication of events or control of manufacturing processes. Can include SPC (Statistical Process Control) or batch sampling schemes.

Strip-down

See Breakdown Inspection.

Subcontractor

Supplier of goods or services on a contract basis usually for the duration of a project.

Supplier

In general, every-day English the person who supplies you with goods or services. [**Note:** Previously this was confusing when reading earlier versions of the ISO 9000 series of standards. They seem to be written on the assumption that

ISO 9001/2:1994 is being called up in a contract by a 'customer' (or 'purchaser' in the 1987 version) and that you as implementing ISO 9001/2 are 'the supplier' to this customer. Hence when reading earlier versions of ISO 9001/2 in most places where it states 'the supplier shall' it is referring to **you**. People who supply you were then called subcontractors or vendors. 9001:2000 has been amended to make it more sensible. See Para 3].

Tally-chart	A simple chart of crosses against each occurrence, giving a rough picture of the process or sample. (See also Histogram.)
Tender	See Contract.
Test certificate	See Certificate of Test.
'TickIT'	A UK approval for ISO 9001 for computer software products or embedded software in electronic goods or control units. As well as checking the DQS against ISO 9001:1994, the system must also comply with the relevant section of ISO 9000–3:1991.
	It is no longer mandatory in UK to have TickIT approval for software. However in practice this does not generally make it less expensive for certification for ISO 9001:2000 for software. The certification body will still have to use a 'software competent' assessor, who tend to be expensive.
Tolerance	The difference between the upper and lower bounds of a specified value.
Toll manufacture	A term sometimes used in the chemical industry for 'bought in and supplied on' (see Chapter 6).
Traceability	The history of a material back to the original melt, batch or lot of material, or the process of grouping components made in uniform conditions or parts into discrete lots or batches. (**Note**: not a requirement of ISO 9001 in itself. Usually carried out because it is a contract requirement, or the product itself is of the type that may run the risk of product withdrawal at a future date.) (See also Certificate of Material Traceability and Certificate of Batch Traceability.)
Type approval	The status given to a design that has been shown by a test (or a series of tests) on a representative model that it meets all the requirements of the product or safety specification and is released as suitable for use.
UKAS	United Kingdom Accreditation Service (used to be called the NACCB). The body responsible to the Government for assessing, surveillance and approval (i.e. accreditation) of the Certification Bodies, e.g. URS, NQA, BSI, LRQA, etc.
Validation	Term introduced in the 1994 issue of ISO 9000. Final tests on the finished product (or service) as a whole, e.g. in-service trials.

Variables	The characteristics of an item that are evaluated against a numerical assessment or scale, e.g. length, tolerances, weights, densities, electrical resistance.
Variation	See 'Concession'.
Vendor	Supplier of goods, services, raw materials on an order-by-order basis.
Vendor rating	A method of ranking or giving a score for different suppliers usually based on number of reject batches, late deliveries and price. Usually only applicable where a company has a large number of suppliers. It can also easily become very misleading or very complex. In actual practice it is usually found that a firm is excellent at supplying 'A' and 'B' and good at supplying 'C', but not as good at supplying 'X' and 'Y' and useless at trying to supply 'Z'.
Verification	An overall check to ensure items are correct. Can also be an inspection of your inspection department and a verification of their findings. In the design department, the checking of a drawing or calculation sheet by an independent person with suitable experience or qualification.
Verification of purchased product	One of the most commonly misinterpreted clauses (7.4.3) of ISO 9001. Often erroneously documented as ONLY 'goods-inward inspection'. It is also necessary to consider the requirements to document and specify where you, as the buyer, wish to examine the goods before despatch or at source. With similar requirements, where appropriate, for your customer to also examine at source if they so wish. (This is a common requirement for MoD, aerospace, clothing or food industry to request that their representative can examine castings, forgings, fabric, cloth, meat e.g. raw materials, before they are processed.)
Vickers test	See Hardness Testing.
Visual reference standard	A controlled example. To give a reference for subjective visual inspection.
Waiver	See Concession.
'Widgets'	A common English term, often used to describe miscellaneous or general standard everyday engineering products.
Work instructions	How an individual is to perform a particular task. See Chapter 5 to see where placed in the hierarchy of documents. These may not exist in the DQS of a small firm, having been absorbed into the procedures to reduce the volume of the DQS and make simpler.
Workmark	A unique rubber stamp or metal punch. Used to show the item has been checked or tested, which also identifies the inspector.

Zero defects   A philosophy of all employees striving to produce absolutely
                                no defective product (or service) in each day. Similar
                                philosophy of everybody attempting to be 'Right First Time'.
                                The philosophy is absolutely right but it must be tempered
                                with the realisation that only approximately 10% of errors
                                are the fault of operatives. 90% of all quality calamities are
                                caused by management by: (i) poor systems and controls,
                                (ii) insufficient information or instructions, (iii) supplying
                                incapable equipment (see Standard Deviation, Normal
                                Distribution and SPC), (iv) lack of adequate training.

Appendix C

9001:2000 vs 9001:1994 tables of the significant changes

TABLE C.1 Significant additional requirements reference 9001:1994

Old 1994 Clause	Requirement	Probable/possible implications for your organisation	New ISO 9001:2000 Clause
Totally new	You must have as minimum documented procedures for **only:** Document Control Control of Records Internal audits Control of Non-conforming product Corrective Actions Preventive Actions	You may decide it appropriate to have only these mandatory procedures. However, do note you must have appropriate 'documents' to demonstrate control of all processes. The author's advice is to have a full set of procedures as previously prescribed under ISO 9001:1994	Various
Totally new	Requires organisation to continually improve the effectiveness of the quality management system	Organisation will need a policy of continuous improvement and have method for implementing that policy	4.1

Old 1994 Clause	Requirement	Probable/possible implications for your organisation	New ISO 9001:2000 Clause
Totally new	Determine the sequence and interaction of identified processes	The organisation will need to have description of the processes and how they relate and interface. (*Author's note:* Some describe effectively in 'words'. However, some certification bodies are insisting on a one page flow diagram. I don't know where it demands this in the standard or why the firm suddenly becomes super effective because of this extra sheet in their Quality Manual)	4.1b
Totally new	Organisation to implement action for continual improvement of all processes	Actions will be needed to show improving the performance of processes (not just manufacturing/service provision, but contract review, design, purchasing etc.)	4.1f
Totally new	Commitment to continually improve the effectiveness of the QMS	Amend Quality Policy to include continually improve QMS with resulting obligations	5.3b
Totally new	Measurable quality objectives are established at relevant functions and levels within the organisation	Quality objectives should be set at relevant levels. (*Note:* This is not specific or restricted to 'Quality' topics. You can include objectives for the business, for the processes, for the product or individuals)	5.4.1
Totally new	Top management to ensure appropriate communication processes are established within the organisation	Processes for communicating information (undefined), up, down and across the organisation are needed	5.5.3
Totally new	Communication takes place regarding the effectiveness of the QMS	Performance of the system and processes regularly communicated to staff	5.5.3

5.6	Performance to meet quality objectives must be collected and reviewed, new objectives and continuous improvement to be actioned and recorded	The Management Review to assess opportunities for improvement	Totally new
5.6.3	Review should result in decision to improve products, processes and the system in terms of the actions required	Outputs from management review to include action, improvement of the QMS in processes, products and resources	Totally new
7.1a	Quality objectives required with regard to possible products or service delivery, improvement	Organisation to determine the quality objectives for the product	Totally new
8.1	You are required to demonstrate how you monitor, measure, analyse and improve your processes. (*Author's note*: Measure implies direct comparison to a physical standard. Monitor is an unspecified check e.g. a judgement)	The standard requires the monitoring, measurement, analysis for processes and systems, including 'continual improvement' processes	Totally New
8.1c	Collect appropriate data, plan, provide resources and implement actions to improve the products, processes and systems	The organisation to analysis results and plan improvement processes that are needed to continually improve the effectiveness of the management system	Totally new
8.2.1	Processes need to be developed and installed to monitor 'customer perceptions'. Must be proactive, just waiting for complaints is not enough	Monitor information relating to customer 'perception' as to whether the organisation has met customer requirements	Totally new
8.4a	Need process to provide objective analysis of customer satisfaction	Analysis of data on customer satisfaction	Totally new
8.5.1	This is a new requirement. The organisation needs to put in place processes to actively find better ways of doing things	Continually improve the effectiveness of the quality management system through the use of the quality policy, quality objectives, audit results, analysis of data, corrective and preventive actions and management reviews	Totally new

Old 1994 Clause	Requirement	Probable/possible implications for your organisation	New ISO 9001:2000 Clause
4.1.1	Documented statements of quality policy and quality objectives	Quality policy to include references to quality objectives	4.2.1a
4.1.1	Top management provide evidence of its commitment to the development and implementation of the quality management system	'Top management' have to demonstrate their involvement in the development and implementation of the QMS	5.1
4.1.1	Top management commitment by establishing quality objectives	'Top management' must have process for establishing, recording and actioning 'quality objectives'	5.1c
4.1.1	Establishing and reviewing quality objectives	Amendments required to Quality Policy and process installed	5.3c
4.1.3	Top management provide evidence of its commitment to continually improving the effectiveness of the quality management system	'Top management' have to demonstrate they continually seek and implement better ways of doing things	5.1
4.1.3	Top management conduct management reviews	'Top management' will need to demonstrate **involvement** in the Quality System Management Reviews	5.1d
4.1.3	Management representatives to ensure awareness of customer requirements is promoted throughout the organisation	Additional new requirements for the Management representative	5.5.2
4.1.3	Records '**from**' Management Reviews to be maintained (previously 'of')	Could be interpreted as requiring not just records of meeting, but records kept to show actions taken and completed	5.6

102

4.1.3	List of items that must be covered at the Management Review	As a minimum (you can expand it) you must have agenda and records covering 5.6.2 items (a) to (g) and 5.6.3 items (a) to (c)	5.6.
4.1.3	Management review inputs to include changes that could affect the quality management system	Review need to cover changes to business e.g., growth or decline, new products etc.	5.6.2
4.1.3b	Analysis of data on conformance to product requirements	Possibly require more correlation of data with product requirements than in the past	8.4b
4.2.1	Quality Manual includes the scope of the QMS. It must also state and justify any exclusions. (*Note.* At para 1.2 you can only exclude activities within clause 7, 'Product Realisation'. You cannot exclude any of the other system requirements)	In the Quality Manual a clear statement is required of what the company actually does that is included within its ISO 9001: 2000 registration. Also what is excluded with justification. (*Note:* You cannot exclude activities such as design. However you may choose not to register all your 'products' or 'services')	4.2.2
4.3	Top management to provide evidence of its commitment to communicating to all staff the importance of meeting regulatory and legal requirements as well as customer requirements	'Top management' demonstrate its commitment by installing processes and communication to achieve the requirements of customers, *regulatory*, *legal* and other defined interested parties	5.1a
4.3	Determine requirements **not** specified by the customer but necessary for specified or intended use, where known	Anticipate customer's needs and expectations or the requirements that are essential for the product to fulfil its intended purpose. (*Author's note:* Be careful that in trying to meet this new requirement, you **do not** become in breach of contract)	7.2.1b
4.3	Determine statutory and regulatory requirements related to the product, or service provided	Must ensure product or service provided is legal and meets all regulatory requirements of the destination country of the product	7.2.1c
4.3	Determine any additional requirements for the product or service provided, **determined by the organisation**	(*Author's note:* be careful that in trying to meet this new requirement, you **do not** become in breach of contract)	7.2.1d

Old 1994 Clause	Requirement	Probable/possible implications for your organisation	New ISO9001:2000 Clause
4.3	Communicating with customers relating to product information	QMS now needs to **include** marketing activities, brochure, adverts etc. as they communicate product information	7.2.3a
4.3	Determine and implement effective arrangements for communicating with customers relating to enquiries	Must describe and have appropriate method of communicating with customers	7.2.3b
4.3.1	Review the **requirements** of an enquiry or a contract	Previously a **contract** review, is now a total review of the total 'requirements'	7.2.2
4.3.2	Top management to ensure customer requirements are determined plus seek to ensure they include all customer needs and expectations	Involvement of 'top management', ensuring the organisation is more pro-active and seek to establish customer needs and expectations before commencing a job. There will need to be a defined process	5.2
4.4	If you 'design' your product or service it **must** be included within your scope of registration, the assessment process and the Documented Quality Management System	Previously many firms who designed their product or service were misleadingly registered to ISO 9002:1994 and left design out of their scope. This is no longer permitted	1.2
4.4.4	The standard requires the design input to include information derived from previous similar designs where applicable	The standard stating the obvious, that was omitted from previous version	7.3.2c
4.4.5	Design and development processes use appropriate information from purchasing, production or service provision	(*Author's note*: One would think this was obvious. However, it is **not** uncommon in large organisations to find 'Design Offices' in an 'Ivory Tower' and never talk to production. Sometimes I have found hand amended drawings on the shop-floor as the design or drawing tolerances or specifications are just not achievable)	7.3.3b

7.3.3	(*Author's note*: I find this omission unbelievable. Consider (i) problems with Nuclear Power Stations, submarines etc. (ii) UK has literally a mountain of old fridges they cannot dispose of!)	No longer required to consider final disposal of product at end of life — 4.4.5
7.3.4	Must have an effective 'design review' **not** just a progress meeting	Design reviews to evaluate the ability of the results of design and development to 'meet requirements' and identify any problems and propose necessary actions — 4.4.6
7.3.6	Previously validation was stated as mandatory, but was not always possible e.g. your 'product' was a written report, or design proposal? 'Where practicable' has now been added	Validation to be completed before delivery 'where practical' — 4.4.8
7.3.7	A more rigorous review of changes is needed. Consider if a recall process required to modify or recall product in service, and held in stock. Are new maintenance instructions required	Review of design development changes to include evaluation of the **effect of the changes** on constituent **parts** and products **already delivered** — 4.4.9
8.4.d	Must have process and show analysis of poor/good suppliers	Analysis of data reference suppliers — 4.6.2
7.4.1	Whilst not required was usually defined in the majority of existing DQS	Requires **criteria** for selection of suppliers to be established — 4.6.2a
7.4.1	Must have records showing evaluation and re-evaluation of suppliers (Note: re-evaluation is a new requirement)	Criteria for **evaluation** and **re-evaluation** of suppliers to be established — 4.6.2b
7.5.4	No change really, just some people could not see this far	Makes clear that Customer's Property can include 'intellectual property', e.g. designs, specifications, confidential information, personal information etc. — 4.7

Old 1994 Clause	Requirement	Probable/possible implications for your organisation	New ISO 9001:2000 Clause
4.9	Manufacturing the product or delivering the service with aim of enhancing customer satisfaction	Focus of the organisation should move beyond simple conformance to requirements, now a focus on customer satisfaction. (*Note:* See also significant new obligations under contract review para 7.2, an attitude of 'we have met your contract specification and don't care if it doesn't meet your needs or expectations' is no longer satisfactory)	5.2
4.9b	Organisation to determine and manage the work environment needed to achieve conformity to product requirements	Not a welfare issue. The environment must not be detrimental to meeting product or service delivery requirements	6.4
4.10	Organisation to define process validation arrangements and method of **re-validation**	Is there an implied requirement that validation should be checked periodically?	7.5.2
4.12	**Delivery** processes or effect to be considered	Delivery process should be included in the QMS, previously stopped at final inspection and pack	7.5.1f
4.12	Organisation to identify the product status with respect to monitoring and measurement requirements	Previously there was a specific clause that required 'inspection status', e.g. awaiting test, quarantine, reject, OK etc. Not as clearly specified but still required	7.5.3
4.14	Analysis of data to provide information on characteristics and trends to provide opportunities for preventive action	Analysing data for control and improvement purposes	8.4c
4.14.2a	Customer **feedback** to be recorded in addition to customer complaints	Customer feedback, both positive and negative, needs to be recorded, reviewed and if necessary actioned	7.2.3c
4.14.2	Review and, if appropriate take action when nonconforming product is detected after despatch	A process is required to review and, if appropriate, to inform customers, users, stockholders, etc. If appropriate action product recall, modification program, replacements etc.	8.3

4.17	Internal audits to also determine whether the quality management system conforms to the requirements of the International Standard	Extra line/activity in the annual audit schedules	8.2.2
4.17	Plan the audit program taking into consideration the status and importance of the processes and areas to be audited including the results of previous audits	The additional requirement for audit planning to take into account importance of the department, product or processes. Also **previous audit results**, e.g. if department or process has many NCRs, audit more often or vice versa	8.2.2
4.18	Personnel performing work affecting product quality to be **'competent'** on the basis of appropriate education, training, skills and experience	If staff are claiming competency on basis of qualifications and formal training you need copies of certificates or confirmation of training in file	6.2.1
4.18	Determine competence requirement of task and allocated personnel	Need simple 'job or task evaluation' process	6.2.1a
4.18	Evaluate, re-evaluate effectiveness of 'training'	There will need to be some process of formal appraisal to confirm and re-confirm competence (e.g. a person can have had training, have a certificate and still be incompetent). Action taken if judged not competent	6.2.2c
4.18	Ensure that its employees are aware of the relevance and importance of their activities	Possible requirements for improved training, handbooks, guidance notes etc. covering knowledge of product and customer requirements	6.2.2d
4.18	Comprehensive records of education, skills, experience	Full records of education, skills and experience required	6.2.2e
4.20	Includes as one method of measurement 'statistical techniques'	(*Note*: Under the 1994 standard, it was common to disclaim Statistical Techniques as 'not applicable'. Do not make the error of trying to disclaim this clause in total as it is not in Clause 7. In the preamble to the standard, [Para 1.2] it states you can only disclaim items in Clause 7)	8.1

TABLE C.2 Significant additional requirements reference ISO 9001:2000

Clause in ISO 9001: 2000	Requirement	Probable/possible implications for your organisation	Clause ISO 9001: 1994
4.1	Requires organisation to continually improve the effectiveness of the quality management system	Organisation will need a policy of continual improvement and have method for implementing that policy	Totally new
4.1b	Determine the sequence and interaction of identified processes	The organisation will need to have description of the processes and how they relate and interface. (*Author's note*: Some describe effectively in 'words'. However, some certification bodies are insisting on a one page flow diagram. I don't know where it demands this in the standard or why the firm suddenly becomes super effective because of this extra sheet in their Quality Manual)	Totally new
4.1f	Organisation to implement action for continual improvement of all processes	Actions will be needed to show improving the performance of processes (not just manufacturing/service provision, but contract review, design, purchasing etc.)	Totally new
4.2.1a	Documented statements of quality policy and quality objectives	Quality policy to include reference to quality objectives are required	4.1.1
4.2.2	Quality Manual includes the scope of the QMS. It must also state and justify any exclusions. (*Note*: At para 1.2 you can only exclude activities within clause 7, 'Product Realisation'. You cannot exclude any of the other system requirements)	In the Quality Manual a clear statement is required of what the company actually does that is included within its ISO 9001:2000 registration. Also what is excluded with justification. (*Note*: You cannot exclude activities such as design. However you may choose not to register all your 'products' or 'services')	4.2.1
5.1	Top management provide evidence of its commitment to the development and implementation of the quality management system	Top management have to demonstrate their involvement in the development and implementation of the QMS	4.1.1

4.1.3	'Top management' have to demonstrate they continually seek and implement better ways of doing things	Top management provide evidence of its commitment to continually improving the effectiveness of the quality management system	5.1
4.3	'Top management' demonstrate its commitment by installing processes and communication to achieve the requirements of customers, *regulatory*, *legal* and other defined interested parties	Top management to provide evidence of its commitment to communicating to all staff the importance of meeting regulatory and legal requirements as well as customer requirements	5.1a
4.1.1	'Top management' must have process for establishing, recording and actioning 'quality objectives'	Top management commitment by establishing quality objectives	5.1c
4.1.3	'Top management' will need to demonstrate **involvement** in the Quality System Management Reviews	Top management conduct management reviews	5.1d
4.3.2	Involvement of 'top management', ensuring the organisation is more pro-active and seek to establish customer needs and expectations before commencing a job. There will need to be a defined process	Top management to ensure customer requirements are determined plus seek to ensure they include all customer needs and expectations	5.2
4.9	Focus of the organisation should move beyond simple conformance to requirements, now a focus on customer satisfaction. (*Note.* See also significant new obligations under contract review para 7.2., an attitude of 'we have met your contract specification and don't care if it doesn't meet your needs or expectations' is no longer satisfactory)	Manufacturing the product or delivering the service with aim of enhancing customer satisfaction	5.2
Totally new	Amend Quality Policy to include continually improve QMS with resulting obligations	Commitment to continually improve the effectiveness of the QMS	5.3b
4.1.1	Amendments required to Quality Policy and process instilled	Establishing and reviewing quality objectives	5.3c

Clause in ISO 9001: 2000	Requirement	Probable/possible implications for your organisation	Clause ISO 9001: 1994
5.4.1	Measurable quality objectives are established at relevant functions and levels within the organisation	Quality objectives should be set at relevant levels. (*Note*: This is not specific or restricted to 'Quality' topics. You can include objectives for the business, for the processes, for the product or individuals)	Totally new
5.5.2	Management representatives to ensure awareness of customer requirements is promoted throughout the organisation	Additional new requirements for the Management representative	4.1.3
5.5.3	Top management to ensure appropriate communication processes are established within the organisation	Processes for communicating information (undefined), up, down and across the organisation are needed	Totally new
5.5.3	Communication takes place regarding the effectiveness of the QMS	Performance of the system and processes regularly communicated to staff	Totally new
5.6	The Management Review to assess opportunities for improvement	Performance to meet quality objectives must be collected and reviewed, new objectives and continual improvement to be actioned and recorded	Totally new
5.6	Records **'from'** Management Reviews to be maintained (previously 'of')	Could be interpreted as requiring not just records of meeting, but records kept to show actions taken and completed	4.1.3
5.6.	List of items that must be covered at the Management Review	As a minimum (you can expand it) you must have agenda and records covering 5.6.2 items (a) to (g) and 5.6.3 items (a) to (c)	4.1.3
5.6.2	Management Review inputs to include changes that could affect the quality management system	Review need to cover changes to business, e.g. growth or decline, new products etc.	4.1.3

Clause	Requirement	Notes	Old ref
5.6.3	Outputs from management review to include action, improvement of the QMS in processes, products and resources	Review should result in decision to improve products, processes and the system in terms of the actions required	Totally new
6.1	Determine and provide the resources to continually improve effectiveness of the QMS	Plan for, or at least identify, resources to improve the QMS	Totally new
6.2.1	Personnel performing work affecting product quality to be '**competent**' on the basis of appropriate education, training, skills and experience	If staff are claiming competency on basis of qualifications and formal training you need copies of certificates or confirmation of training in file	4.18
6.2.1a	Determine competence requirement of task and allocated personnel	Need simple 'job or task evaluation' process	4.18
6.2.2c	Evaluate, re-evaluate effectiveness of 'training'	There will need to be some process of formal appraisal to confirm and re-confirm competence (e.g. a person can have had training, have a certificate and still be incompetent). Action taken if judged not competent	4.18
6.2.2d	Ensure that its employees are aware of the relevance and importance of their activities	Possible requirements for improved training, handbooks, guidance notes etc. covering knowledge of product and customer requirements	4.18
6.2.2e	Comprehensive records of education, skills, experience	Full records of education, skills and experience required	4.18
6.4	Organisation to determine and manage the work environment needed to achieve conformity to product requirements	Not a welfare issue. The environment must not be detrimental to meeting product or service delivery requirements	4.9b
7.1a	Organisation to determine the quality objectives for the product	Quality objectives required with regard to possible products or service delivery, improvement	Totally new

Clause in ISO 9001: 2000	Requirement	Probable/possible implications for your organisation	Clause ISO 9001: 1994
7.2.1b	Determine requirements **not** specified by the customer but necessary for specified or intended use, where known	Anticipate customer's needs and expectations or the requirements that are essential for the product to fulfil its intended purpose. (*Author's note.* Be careful that in trying to meet this new requirement, you **do not** become in breach of contract)	4.3
7.2.1c	Determine statutory and regulatory requirements related to the product, or service provided	Must ensure product or service provided is legal and meets all regulatory requirements of the destination country of the product	4.3
7.2.1d	Determine any additional requirements for the product or service provided, **determined by the organisation**	(*Author's note.* Be careful that in trying to meet this new requirement, you **do not** become in breach of contract)	4.3
7.2.2	Review the **requirements** of an enquiry or contract	Previously a **contract** review, is now a total review of the total 'requirements'	4.3.1
7.2.3a	Communicating with customers relating to product information	QMS now needs to **include** marketing activities, brochure, adverts etc. as they communicate product information	4.3
7.2.3b	Determine and implement effective arrangements for communicating with customers relating to enquiries	Must describe and have appropriate method of communicating with customers	4.3
7.2.3c	Customer **feedback** to be recorded in addition to customer complaints	Customer feedback, both positive and negative, need to be recorded, reviewed and if necessary actioned	4.14.2a
7.3.2c	The standard requires the design input to include information derived from previous similar designs where applicable	The standard stating the obvious, that was omitted from previous version	4.4.4

4.4.5	(*Author's note.* One would think this was obvious. However, it is **not** uncommon in large organisations to find 'Design Offices' in an 'Ivory Tower' and never talk to production. Sometimes find hand amended drawings on the shop-floor as drawing tolerances or specification are just not achievable)	Design and development processes use appropriate information from purchasing, production or service provision	7.3.3b
4.4.5	(*Author's note:* I find this omission unbelievable. Consider (i) problems with Nuclear Power Stations, submarines etc., (ii) UK has literally a mountain of old fridges they cannot dispose of!)	No longer required to consider final disposal of product at end of life	7.3.3
4.4.6	Must have an effective 'design review' **not** just a progress meeting	Design reviews to evaluate the ability of the results of design and development to 'meet requirements' and identify any problems and propose necessary actions	7.3.4
4.4.8	Previously validation was stated as mandatory, but was not always possible e.g. your 'product' was a written report, or design proposal? 'Where practicable' has now been added	Validation to be completed before delivery 'where practical'	7.3.6
4.4.9	A more rigorous review of changes is needed. Consider if a recall process required to modify or recall product in service, and held in stock. Are new maintenance instructions required	Review of design development changes to include evaluation of the **effect of the changes** on constituent **parts** and products **already delivered**	7.3.7
4.6.2a	Whilst not required was usually defined in the majority of existing DQS (Note: Re-evaluation is a new requirement)	Requires **criteria** for selection of suppliers to be established	7.4.1
4.6.2b	Must have records showing evaluation and re-evaluation of suppliers	Criteria for **evaluation** and **re-evaluation** of suppliers to be established	7.4.1
4.12	Delivery process should be included in the QMS, previously stopped at final inspection and pack	**Delivery** processes or effect to be considered	7.5.1f

Clause in ISO 9001: 2000	Requirement	Probable/possible implications for your organisation	Clause ISO 9001: 1994
7.5.2	Organisation to define process validation arrangements and method of **re-validation**	Is there an implied requirement that validation should be checked periodically?	4.10
7.5.3	Organisation to identify the product status with respect to monitoring and measurement requirements	Previously there was a specific clause that required 'inspection status', e.g. awaiting test, quarantine, reject, OK etc. Not as clearly specified but still required	4.12
7.5.4	Makes clear that Customer's Property can include 'intellectual property', e.g. designs, specifications, confidential information, personal information etc.	No change really, just some people could not see this far	4.7
8.1	The standard requires the monitoring, measurement, **analysis** for processes and systems, including 'continual improvement' processes	You are required to demonstrate how you monitor, measure, analyse and improve your processes. (*Author's note*: Measure implies direct comparison to a physical standard. Monitor is an unspecified check, e.g. a judgement)	Totally New
8.1	Includes as one method of measurement 'statistical techniques'	(*Note*: Under the 1994 standard, it was common to disclaim Statistical Techniques as 'not applicable'. Do not make the error of trying to disclaim this clause in total as it is not in Clause 7. In the preamble to the standard, [Para 1.2] it states you can only disclaim items in Clause 7)	4.20
8.1c	The organisation to plan and implement the analysis and improvement processes needed to continually improve the effectiveness of the management system	Plan, provide resources and implement actions to improve the products, processes and systems	Totally new

Clause	Requirement	Notes	Old ref
8.2.1	Monitor information relating to customer 'perception' as to whether the organisation has met customer requirements	Processes need to be developed and installed to monitor 'customer perceptions'. Must be proactive, just waiting for complaints is not enough	Totally new
8.2.2	Internal audits to also determine whether the quality management system conforms to the requirements of the International Standard	Extra line/activity in the annual audit schedules	4.17
8.2.2	Plan the audit program taking into consideration the status and importance of the processes and areas to be audited including the results of previous audits	The additional requirement for audit planning to take into account importance of the department, product or processes. Also **previous audit results**, e.g. if department or process has many NCRs, audit more often or vice versa	4.17
8.3	Review and, if appropriate take action when nonconforming product is detected after despatch	A process is required to review and, if appropriate, to inform customers, users, stockholders, etc. If appropriate action product recall, modification program, replacements etc.	4.14.2
8.4a	Analysis of data on customer satisfaction	Need process to provide objective analysis of customer satisfaction	Totally new
8.4b	Analysis of data on conformance to product requirements	Possibly require more correlation of data with product requirements than in the past	4.1.3b
8.4c	Analysis of data to provide information on characteristics and trends to provide opportunities for preventive action	Analysing data for control and improvement purposes	4.14
8.4.d	Analysis of data reference suppliers	Must have process and show analysis of poor/good suppliers	4.6.2
8.5.1	Continually improve the effectiveness of the quality management system through the use of the quality policy, quality objectives, audit results, analysis of data, corrective and preventive actions and management reviews	This is a new requirement. The organisation needs to put in place processes to actively find better ways of doing things	Totally new

Appendix D
Draft Quality Manual

On the following pages there is a **draft model**, template or workbook to assist you prepare your very own:

'QUALITY POLICY MANUAL'

Notes:

1.   Throughout this draft or template there will be at appropriate places, possible alternatives or additional text for you to consider. These may be more applicable to your organisational structure, management style, products or services.

These alternatives are shown in brackets in the form (**OR** . . . .).

2.   Where the text shows . . . . plc, these are appropriate places for you to insert the name of your own organisation.

THE QUALITY POLICY MANUAL OF

. . . . plc

This is copy number . . . .

Issued to . . . .

This is a controlled (OR uncontrolled) copy

Authorised and signed by . . . .
(Managing Director)

Circ.:   Managing Director
          Work Manager
          Quality Manager (Master Copy)

          . . . .
          . . . .
          . . . .

Page 1 of 24
Issue . . . .   Date . . . .

Contents

NOTE The reference numbers and the titles of the paragraphs of the draft 'Quality Policy Manual' generally correspond with the paragraph, or clause, numbering of ISO 9001:2000. This arrangement makes it easy for an auditor to check.

NOTE For simplicity, the Quality Policy Manual and Operating Procedures will refer to BS EN ISO 9001:2000, simply as ISO 9001.

0.1   INTRODUCTION TO THE ORGANISATION (OR COMPANY)

.............plc was founded in 19... Its current Managing Director is
.... who took over the responsibility in 19..., previously s/he was at
....

The executive responsible for Quality is the Managing Director (**OR** the Design and Development Director), (**OR** the ....Director), (**OR** the Senior Partner), (**OR** the Managing Director, whilst maintaining executive control, review and responsibility has delegated to the Quality Assurance Manager the day-to-day management and reporting of the Quality System), (**OR** the Managing Director (**OR** ....) who, whilst maintaining executive control, review and responsibility, has delegated the day-to-day management and reporting of the Quality System to .... Consultancy. This is done formally through an annual purchase order specifying those duties, with a monthly (**OR** 3-monthly) written report which is formally reviewed by the firm's executive with responsibility for quality.

The company's products (**OR** services) which it supplies (**OR** manufactures), (**OR** processes), (**OR** stores), (**OR** transports) (**OR** ....) are ...., and ....

These are provided, using trained and experienced staff, by the process of ....
....

The company operates from a modern unit factory on one site (**OR** detail others) which is in (**OR** adjacent to) the town of .... in the county (**OR** state) of .... This is conveniently adjacent to the M ... and M ... motorways (**OR** A ... main roads) (**OR** .... Airport) (**OR** .... Deep Water Port) (**OR** .... Railway).

..............plc's ISO 9001:2000 scope is .... We have excluded .... from the scope of registration and the documented control, because ....

(**OR**
.... plc's ISO 9001:2000 scope is 'Manufacture of metal components and sub-assemblies'. We have excluded 'Design' from the scope of registration and the documented controls, because we always make to customers' designs or specifications.)

(**OR**
.... plc's ISO 9001:2000 scope is 'Supply and installation of fitted carpets'. we have excluded 'Design' from the scope of registration and the documented controls, because we do not design at all. We have also excluded 'calibration' as we have no instruments that have to be calibrated to national, or manufacturer's standards. We use tapes and rules of good quality. These are of high grade but are only used for comparison purposes, as carpets are cut oversize 'off the roll' and then cut to fit.)

4 QUALITY MANAGEMENT SYSTEM

4.1 General requirements

. . . . plc have developed, written and formally issued, installed, maintained and, where possible, continue to improve our quality management system based on the requirements of ISO 9001:2000.

. . . . plc have:

(a) identified the processes needed for the quality management system and its application throughout the organisation;

(b) determined the sequence and interaction of these processes;

(c) determined the criteria and methods needed to ensure that both the operation and control of these processes are effective;

(d) made available the resources and information necessary to support the operation and monitoring of these processes;

(e) set up systems to monitor, measure and analyse these processes;

(f) taken and continue to take any actions necessary to achieve planned results and continual improvement of these processes.

These processes are managed by . . . . plc in accordance with the requirements of ISO 9001:2000.

. . . . plc purchase appropriate raw material, components or services. However none of the 'processes' are sub-contracted to a third party.

(**OR** . . . . plc find it appropriate to purchase from a third party 'processes', e.g Design (**OR** Galvanising, **OR** . . . .) that affects product or service conformity with requirements, ... plc ensures it retains effective control over such processes. Control of such processes that are from a source outside the organisation, will be identified within the quality management system.)

4.2 Requirements for Documentation

4.2.1 General Requirements for Documentation

The quality management system documentation has been developed, written, issued, installed and maintained, includes the following:

(a) a quality policy and quality objectives statement, which is formally issued, signed and dated;

(b) a quality policy manual;

(c) operating procedures;

(d) any additional documents or forms needed by . . . . plc to ensure the effective planning, operation and control of its processes;

(e) records required to demonstrate effective control or protect . . . . plc or, any records required by ISO 9001:2000, are established and maintained.

4.2.2 Quality Policy Manual

. . . . plc have established and maintain a quality policy manual that includes

(a) the scope of the quality management system, including details of and justification for any exclusions are in Section 0.1;

(b) reference to the associated operating procedures, see Annex 'B';

(c) a description of the interaction between the processes of the quality management system, see Annex 'C';

4.2.3 Control of documents

Documents required by the quality management system are controlled.

Documented procedures have been developed and installed to provide the necessary controls to

(a) check documents to ensure they are correct and also to formally approve before issue;

(b) review and update documents where necessary and to re-approve and re-issue such new or amended documents;

(c) ensure that changes are recorded and the current revision status of documents are identified;

(d) ensure that relevant versions of appropriate documents are readily available at the points of use;

(e) ensure that documents remain legible and readily identifiable;

(f) ensure that appropriate documents of external sources are identified and their distribution controlled;

(g) prevent the unintended use of obsolete documents, and to apply suitable identification to them if they are retained for historical reference or any purpose.

4.2.4 Control of records

Records are established and maintained to provide evidence of conformity to requirements and of the effective operation of the quality management system. Controls exist to ensure records remain legible, readily identifiable and retrievable. A documented procedure has been established to define the controls needed for the identification, storage, protection, retrieval, retention time and disposition of records.

5 MANAGEMENT RESPONSIBILITY

5.1 Management commitment

The overall management structure is shown at Annex 'A'. Top management provides evidence of its commitment to the development and implementation of the quality management system and continually improving its effectiveness by:

(a) communicating to all of .... plc staff, the importance of meeting customer requirements, whilst also meeting statutory and regulatory requirements;

(b) documenting and authorising the quality policy;

(c) ensuring that quality objectives are established;

(d) conducting management reviews;

(e) ensuring the availability of resources.

5.2 Focusing on customer requirements

Top management continually ensures that customer requirements are determined and are met with the aim of enhancing customer satisfaction.

5.3 Quality policy

The following is the formal Quality Policy statement of .... plc, it is reviewed regularly to ensure that it remains appropriate for our products and services.

.... plc is committed to comply with requirements of ISO 9001:2000 and to continually improve the effectiveness of the quality management system,

.... plc is committed to seek to continually improve the quality of its products and services. We will always meet the specified requirements and seek to satisfy or exceed the customers' expectations.

The top management of .... plc is fully committed to the Documented Quality Management system. It must be clearly understood that this Quality Policy, The Quality Policy Manual and associated Operating Procedures and systems are mandatory on all staff.

.... plc have introduced systems that will set and review measurable quality objectives. Top management will provide any resources required and with all our staff we will try our best to meet and surpass these objectives.

*A copy of the latest issue of this Quality Policy is displayed in the foyer, the workshop (**OR** ...., ...)*

Signed .... Date ....

(Managing Director)

(**OR** suitable representative of Top Management)

5.4 *Planning*

5.4.1 Quality objectives

Top management have introduced a formal Operating Procedure on Continuous Improvement and Quality Objectives.

These objectives can be on either, the management systems or, our product and services. The objectives will be specific, measurable and consistent with the quality policy.

5.4.2 Quality management system planning

Top management have ensured that

(a)   planning of the quality management system is carried out in order to meet the general and specific requirements of ISO 9001:2000 including the need for continuous improvement and establishing and meeting quality objectives;

(b)   the integrity of the quality management system is maintained when changes to the quality management system are planned and implemented.

5.5 *Responsibility, authority and communication*

5.5.1 Responsibility and authority

Top management have ensured that responsibilities and authorities are defined and communicated within . . . . plc.

The management structure of . . . . plc is shown at Annex 'A' to this Quality Policy Manual. This chart simply shows functional relationships and responsibilities. It does not imply relative seniority or importance of position.

The . . . . in addition to his/her normal responsibilities as a senior executive of the company is the nominated management representative with responsibility for establishing, implementing and maintaining the Quality System and reporting its effectiveness. S/he has full authority to ensure that the system is understood and established throughout the organisation.

(**OR** the . . . . who, whilst maintaining executive control, review and responsibility, has delegated the Quality Assurance Manager with day-to-day management and reporting of the Quality System.)

(**OR** the . . . . who, whilst maintaining executive control, review and responsibility, has delegated the day-to-day management and report of the Quality System to . . . . Consultancy. This is done formally through an annual purchase order specifying their duties, with a monthly (**OR** 3-monthly) written report which is formally reviewed and signed off and dated by the firm's executive with responsibility for quality.)

(**OR** . . . . plc is a small company. The designated management representative for all Quality Assurance matters is the Managing Director.)

(b)   Responsibilities and authority

> The Directors and all managers (**OR** the two partners) and supervisors ensure that all the requirements of the Quality Policy Manual and the Operating Procedures have been fully implemented and are maintained. They also ensure all staff understand the requirements of the Operating Procedures (**OR** Standing or Temporary Work Instructions) affecting their tasks and the requirements of each contract. The Directors, managers and supervisors also ensure that their staff have the necessary procedures, work instructions, training, specifications, drawings, tools and equipment to effectively carry out the work.

> Each employee of the company is responsible for maintaining the specified standards of work for each contract at all times. As a general policy any supervisor may perform the tasks of those under their supervision, if supervisors can demonstrate adequate qualifications and/or experience.

The following staff's responsibilities, authority and inter-relationship are defined and documented below. Other staff's responsibilities are adequately covered by the Operating Procedures (**OR** any specific Standing or Temporary Work Instructions raised).

Managing Director

- Business policy, planning and marketing strategy.
- Marketing and selling activities.
- Overall direction and management of the business.
- Reviews all enquiries or contracts with a value of £(**OR$**) . . . . and above, before submission of an offer or tender and/or on receipt of contract.
- Overall authority on Quality Assurance and ensuring the organisation meets its Health and Safety and other legislative obligations.
- Reviews and approves all major projects, capital purchases and leases.
- Reviews and approves all contract/process related purchase orders above a value of £(**OR$**) . . . .
- etc.

Financial Director

- All accounting and financial affairs and records.
- Management of the administration, personnel and office services.
- etc.

(*Author's Note*: I have seen several small firms where the Financial Director has been designated the QA Manager. These all appear to have worked very well. The Financial Director tends to be independent of the processes and also is used to the discipline of financial auditing. It also gives him an excellent insight into what is actually going on!)

Production Manager

- Reviews and approves all contracts below a value of £(**OR$**) . . . . . ., before submission of an offer or tender and/or on receipt of contract.
- Assisting Managing Director as required on major contract reviews.
- Direct liaison with customers on technical or contract related matters.
- Ensuring items and services are supplied exactly to specification and within cost and delivery targets.
- Control of contract planning office (**OR** . . . . office) ensuring the filing systems and control of documents meet the requirements of the Operating Procedures.

- Identifying training needs for his/her staff, arranging training and maintaining training records.
- Raising and authorising purchase orders of, or below a value of £(**OR$**) . . . .
- Control of raw materials and components stock.
- Planning and scheduling of workload, ensuring the contract requirements are understood by operational staff.
- Ensuring all work is given to personnel who have adequate training and, as necessary, it is adequately covered by appropriate instructions, drawings, specifications, etc.
- Oversees all operational and day-to-day activities.
- Ensuring all plant is given suitable maintenance to prescribed and documented procedures.
- etc.

Administration and Office Manager

- Responsible for administration and office services.
- Responsible for the overall control of purchasing.
- etc.

. . . . Manager

- etc.

(**OR** . . . . plc is a small firm. The duties above are those normally carried out by the designated person. However, it is necessary for other directors (**OR** partners) to be able to interchange and each are capable of doing each others jobs if the need arises).

(OR 5.5.2 Management Representative

Note: in the absence of the Management Representative these duties revert back to the Managing Director (**OR** the executive responsible for quality), (**OR** . . . . plc is a small firm, and the Managing Director undertakes the following duties as the Management Representative). Whilst it is the responsibility of all of . . . . plc staff to meet quality and contractual requirements, the Management Representative has specific responsibility for:)

5.5.2 Management representative

Top management have appointed . . . . who, irrespective of other responsibilities, who has specific responsibilities and delegated authority that includes

(a) ensuring that processes needed for the quality management system are established, implemented and maintained;

(b) reporting to top management on the performance of the quality management system and any need for improvement, and;

(c) ensuring the promotion of awareness of customer requirements throughout . . . . plc;

(d) responsibility of liaison with external parties on matters relating to the quality management system.

5.5.3 Internal communication

Top management have developed and issued a formal procedure on Internal and External Communication.

5.6   *Management review*

Top management formally reviews . . . . plc's Quality Management System, at planned intervals, to ensure its continuing suitability, adequacy and effectiveness. This review is formally recorded and includes assessing opportunities for improvement and the need for changes to the quality management system, including the quality policy and quality objectives.

This is covered by a designated Operating Procedure.

6   RESOURCE MANAGEMENT

6.1 *Provision of resources*

. . . . plc have determined, planned and provided the resources needed to implement and maintain the quality management system, to continually improve its effectiveness, and to enhance customer satisfaction by meeting customer requirements and expectations.

6.2   *Human resources*

6.2.1   *General*

. . . . plc have procedures to ensure that personnel performing work affecting product or service quality are competent on the basis of appropriate education, training, skills and experience.

6.2.2 *Competence, awareness and training*

. . . . plc have procedures to

(a)   determine the necessary competence for personnel performing work affecting product or service quality;

(b)   provide training or other appropriate actions to satisfy these needs;

(c)   review and evaluate the effectiveness of the actions taken;

(d)   ensure that its personnel are aware of the relevance and importance of their activities and how they contribute to the achievement of the quality objectives;

(e)   maintain appropriate records of education, training, skills and experience.

6.3   *Infrastructure*

. . . . plc plans, provides and maintains the infrastructure needed to achieve conformity to product or service requirements. Including, where applicable

(a)   buildings, workspace and associated utilities;

(b)   process equipment (both hardware and software);

(c)   supporting services (such as transport or communication).

6.4   *Work environment*

. . . . plc continually reviews and manages the work environment needed to achieve conformity to product or service requirements.

7 PLANNING TO REALISE (OR 'MAKE TO HAPPEN') THE PRODUCT AND/OR SERVICE

7.1 *Planning and management of processes directly relating to the product and/or services*

.... plc have developed and implemented procedures and provided the necessary resources, for all the processes to ensure successful realisation of the product or services to conform to specified requirements. Quality objectives to improve the product will be programmed where appropriate.

If appropriate for a particular product or service .... plc may develop a detailed formally documented and controlled 'Plan' going through all the steps required, i.e. from initial conception to final test and delivery.

If such a plan is requested from a customer, we will ask them for their preferred format, as they can be presented in many ways, e.g. tabular, flowchart, pictorial, manuscript, etc.

7.2 *Customer-related processes*

7.2.1 *Determination of requirements related to the product or service*

(a)   .... plc will determine customer requirements, preferably as a clear specification, including any requirements for delivery and post-delivery activities;

(b)   .... plc will also consider requirements not stated by the customer but thought may be necessary for specified or intended use (where the intended use is in fact known);

(c)   .... plc will determine statutory and regulatory requirements related to the product or service;

(d)   .... plc will determine any additional requirements that may enhance the product or service.

If consideration of b, c, or d, above appears to conflict with 'a' (e.g. the customer's specification) the customer is advised formally, to confirm if they wish to amend their specification. Such advice and any subsequent amendments to specification or order requirements are formally reviewed and recorded.

7.2.2 *Review of requirements and the contract related to the product or service*

.... plc will review the requirements related to the product or service. This review will be conducted prior to .... plc's commitment to supply a product or service to the customer (e.g. at submission of tenders, at acceptance of contracts or orders, at acceptance of changes to contracts or orders) and will ensure that

(a)   product or service requirements are adequately defined;

(b)   any differences or inconsistency in contract or order requirements from those in the tender, offer or from those previously expressed are resolved, and;

(c)   .... plc has the ability to meet the defined requirements.

Records of the results of the contract review and actions arising from the review are maintained.

Where the customer provides no documented statement of requirement, the customer requirements will be confirmed and documented internally by .... plc before acceptance.

Where product or service requirements are changed, ... plc will ensure that relevant documents are amended and also that relevant personnel are made aware of the changed requirements.

The customer may order an 'off-the-shelf' or catalogue item. In these cases, the review will be a check to ensure that, there are no additional, unusual or additional requirements.

7.2.3 Customer communication

. . . . plc has determined and implemented effective arrangements for communicating with customers in relation to

(a)   product or service information;

(b)   enquiries, contracts or order handling, including amendments, and;

(c)   customer feedback, including customer complaints.

7.3   Design and development

(**OR** . . . . plc do not 'design' see para 0.1, therefore this clause 7.3 does not apply.)

7.3.1 Design and development planning

. . . . plc will plan and control the design and development of product or service.

During the design and development planning, ... plc will determine

(a)   the design and development stages;

(b)   the review, verification and validation that are appropriate to each design and development stage, and;

(c)   the responsibilities and authorities for design and development.

. . . . plc will manage the interfaces between different groups involved in design and development to ensure effective communication and clear assignment of responsibility.

Plans will be updated, as appropriate, as the design and development progresses.

7.3.2 Input into the Design and development process

Inputs relating to product or service requirements will be determined and records maintained. These inputs will include

(a)   functional and performance requirements;

(b)   applicable statutory and regulatory requirements;

(c)   information derived from previous similar designs, where applicable;

(d)   any other requirements essential for design and development.

These inputs will be reviewed for adequacy. Requirements will be complete, unambiguous and not in conflict with each other.

7.3.3 Output from design and development

The outputs of design and development will be provided in a form that enables verification against the design and development input and will be approved prior to release.

Design and development outputs will

(a)   meet the input requirements for design and development;

(b)   provide appropriate information for purchasing, production and for service provision;

(c)   contain or reference product acceptance criteria or service acceptance criteria, and;

(d)   specify the characteristics of the product or service that are essential for its safe and proper use.

(**OR** (e) provide customer advice on disposition or re-cycling at the end of the product's life. If appropriate, provide advice in case of catastrophic failure, accident, collision, etc.)

7.3.4 Design and development review

At suitable stages, systematic reviews of design and development will be performed in accordance with planned arrangements. The purpose is:

(a)   to evaluate, or check the ability of the results of design and development to meet specified requirements, and;

(b)   to identify any problems and propose necessary actions.

Participants in such reviews will include representatives of functions concerned with the design and development stage(s) being reviewed. Records of the results of the reviews and any necessary actions will be maintained

7.3.5 Verification of design and development

Verification (e.g. drawing or specification checks, etc.) will be performed in accordance with planned arrangements to ensure that the *design and development 'OUTPUTS' meet the specified design and development 'INPUT' requirements.* Records of the results of the verification and any necessary actions will be maintained.

7.3.6 Validation of design and development

Design and development validation (e.g. field trials, product laboratory tests, etc.) will be performed in accordance with planned arrangements to ensure that the *resulting 'product or service' is capable of meeting the requirements for the 'ACTUAL SPECIFIED APPLICATION' or 'INTENDED USE'*, (where the intended use is known).

Wherever practicable, validation will be completed prior to the delivery or implementation of the product. Records of the results of validation and any necessary actions will be maintained.

7.3.7 Control of changes during design and development

Design and development changes will be identified and records maintained. The changes will be reviewed, verified and validated, as appropriate, and approved before implementation.
The review of design and development changes will also include an evaluation of the effect of the changes on product or service already delivered and/or any constituent parts or spares that may be held in stock, or work in progress.

Records of the results of the review of changes and any necessary actions will be maintained.

7.4 Purchasing

7.4.1 Purchasing process

. . . . plc will ensure that purchased items, material or service conforms to specified purchase requirements. The type and extent of control applied to the supplier and the purchased items, material or service will be dependent upon the effect of the purchased items, material or service on subsequent final product provided to the customer.

. . . . plc have and will evaluate and select suppliers based on their ability to supply product or service in accordance with . . . . plc's requirements.
Criteria for selection, evaluation and re-evaluation have been established.
Records of the results of evaluations and any necessary actions arising from the evaluation are maintained.

7.4.2 Information required for purchasing

Purchasing information will clearly and uniquely describe the product or service to be purchased, including if appropriate

(a)   any additional requirements for 'approvals' or in-process or final inspection of product or service, procedures, processes and equipment prior to delivery;

(b)   any additional requirements for qualification of personnel;

(c)   any essential or additional quality management system requirements.

. . . . plc will formally review the specified purchase requirements to ensure the adequacy, prior to their communication to the supplier.

7.4.3 Verification of purchased product or service

. . . . plc have established and implemented inspection, checks or other activities necessary to ensuring that purchased product or service meets specified purchase requirements.

Where . . . . plc or our customer intends to perform inspection or verification at the supplier's premises, or at source . . . . plc will state the intended inspection or verification arrangements and method of product or service release in the purchasing information.

7.5 Provision for Production and/or the Service required

7.5.1 Control of production and service provision

. . . . plc have and will continue to plan and carry out production and service provision under controlled conditions. Controlled conditions will include, as applicable

(a)   information available that describes the characteristics of the product or service required;

(b)   where necessary, suitable operating procedures or work instructions available where the operation actually takes place;

(c)   provision and use of suitable equipment;

(d)   provision of monitoring and measuring devices and instructions when to use and the specified tolerances or process control data;

(e)   appropriate monitoring and measurement;

(f)   suitable release, delivery and post-delivery activities.

7.5.2 Validation of special production and service processes

.... plc will validate any processes for production and service provision that are special in the sense that the resulting output cannot be verified easily or inspected by subsequent monitoring or measurement.

This includes any processes where deficiencies become apparent only after the product is in use or the service has been delivered.

Validation will demonstrate the ability of these processes to achieve the planned results.

.... plc have established arrangements for these processes including, as applicable

(a)   defined criteria for review and approval of the processes;

(b)   approval of capable equipment and qualification of personnel;

(c)   use of specific methods, procedures, instructions or in-process checks or inspections;

(d)   appropriate records will be kept of the development and approval of the validation processes; also appropriate records will be kept of the readings or inspection results arising;

(e)   if appropriate the processes will be subject to revalidation.

7.5.3 Identification and traceability

Where appropriate, ... plc will identify the product or service provided by suitable means throughout product or service realisation.

As necessary, ... plc will identify the status of the product or service with respect to its monitoring and measurement requirements, or if it is awaiting inspection and approval, quarantined, rejected, etc.

Where traceability is a requirement of the contract, ... plc will capture (or create), control and record the unique identification of the product or service and provide the necessary traceability to raw material, suppliers, process, equipment, personnel, project files, etc.

7.5.4 Customers' property

.... plc will exercise care with customers' property while it is under our control or being used by our staff or sub-contractors.

.... plc will identify, verify, protect and safeguard customers' property that has been provided for use in the processes or for incorporation into the product or service to be provided.

If any customers' property is lost, damaged or otherwise found to be unsuitable for use, this will be reported to the customer and records maintained.

. . . . plc staff are made aware that customers' property can include intellectual property such as designs, drawings, specifications, commercial in-confidence information etc.

Where a customer wishes to perform inspection or verification at . . . . plc or a supplier's premises, special arrangements or care may be necessary to prevent one customer viewing a competitor's order or requirements.

7.5.5 *Preservation of product*

. . . . plc will preserve the conformity of product (and constituent parts) during internal processing and delivery to the intended destination.
This preservation will include appropriate identification, handling, packaging, storage and protection.

7.6 *Control of monitoring and measuring devices, including calibration*

. . . . plc will determine the monitoring and measurement to be undertaken and appropriate devices needed to provide evidence of conformity of product or service to specified requirements.

. . . . plc will establish processes to ensure that appropriate monitoring and measurement can be carried out to meet the necessary requirements.

To ensure valid results, where appropriate measuring equipment will

(a) be calibrated or verified at specified intervals, or prior to use, against measurement standards traceable to international or national measurement standards; where no such standards exist, the basis used for calibration or verification will be recorded;

(b) be adjusted or re-adjusted as necessary;

(c) be identified to enable the calibration status to be determined;

(d) be safeguarded from adjustments that would invalidate the measurement result;

(e) be protected from damage and deterioration during handling, maintenance and storage.

In addition, when an equipment is found not to conform to requirements . . . . plc will assess and record the validity of the previous measuring results taken on that equipment. If appropriate . . . . plc will take action on the equipment. Also, if appropriate . . . . plc will take action on any previously supplied product or service that has been affected.

Records of the results of calibration and verification will be maintained

Any computer software used in the monitoring and measurement of appropriate specified requirements will be checked and validated prior to use and if deemed necessary reconfirmed at appropriate intervals.

8 MEASUREMENT, ANALYSIS AND IMPROVEMENT

8.1 General

. . . . plc will plan and implement the monitoring, measurement, analysis and improvement processes needed to:

(a)   demonstrate conformity of the product or service;

(b)   ensure compliance of the quality management system, and;

(c)   continually improve the effectiveness of the quality management system.

This will include determination of applicable methods, including statistical techniques where applicable, and the extent of their use.

8.2 Monitoring and measurement

8.2.1 Customer satisfaction

. . . . plc monitor information relating to customer perception as to whether . . . . plc are meeting customer requirements. The methods for obtaining and using this information is described in the operating procedures.

8.2.2 Internal audits

. . . . plc have implemented documented procedures to carry out internal audits at planned intervals to determine whether the quality management system conforms:

(a)   to the planned arrangements;

(b)   to the requirements of ISO 9001:2000;

(c)   to the quality management system requirements established by . . . . plc is effectively implemented and maintained.

An audit programme will be planned each year, taking into consideration the status and importance of the processes and areas to be audited, as well as the results of previous audits. The audit criteria, scope, frequency and methods will be defined. Selection of auditors and conduct of audits will ensure objectivity and impartiality of the audit process.

Auditors will not audit their own work.

The responsibilities and requirements for planning and conducting audits, and for reporting results and maintaining records is defined in the documented procedure.

The management responsible for the area being audited will ensure that actions are taken without undue delay to eliminate any non-conformity found and the causes. Follow-up activities will include the verification of the actions taken with appropriate recording and reporting.

8.2.3 Monitoring and measurement of processes

. . . . plc apply suitable methods for monitoring and, where applicable, measurement of the quality management system processes. These methods demonstrate the ability of the processes to achieve planned results. When planned results are not achieved, corrective action is taken, as appropriate, to ensure conformity of the product or service.

8.2.4 Monitoring and measurement of product or service

. . . . plc will monitor and measure the characteristics of the product or service to verify that requirements have been met. This is carried out at appropriate stages of the product or service realisation process in accordance with the documented procedures and planned arrangements.

Evidence of conformity with the acceptance criteria are maintained. Records will indicate the person(s) authorising release of product or service.

Only in exceptional circumstances, will product be released and service be delivered before the planned arrangements have been satisfactorily completed. This will always require formal approval, if necessary by the customer.

8.3 *Control of non-conforming product or service*

. . . . plc will ensure that product or services that do not conform to requirements are identified and controlled to prevent unintended use or delivery. The controls and related responsibilities and authorities for dealing with non-conforming product or service are controlled by the documented procedures.

. . . . plc will deal with non-conforming product or service by one or more of the following ways

(a)   by taking action to eliminate the detected non-conformity and its cause;

(b)   by authorising its re-work to make it conform to specification;

(c)   release or acceptance 'as it is' or without additional work under concession by a relevant authority and, where applicable, by customer;

(d)   by repair, or additional processes, which will make it usable but not exactly to specification. This will require release or acceptance under concession by a relevant authority and, where applicable, by the customer;

(e)   by taking action to preclude its original intended use or application e.g. scrap, re-grade and identify as 'seconds', use as raw material for another process, etc.

Records of the nature of non-conformities and any subsequent actions taken, including concessions approved and rejected are maintained.

When non-conforming product or service is corrected, it will be subject to re-verification to demonstrate conformity to the requirements.

When non-conforming product or service is detected after delivery or use has started, ... plc will take action appropriate to the effects, or potential effects, of the non-conformity.

8.4   Analysis of data

. . . . plc will determine, collect and analyse appropriate data to demonstrate the suitability and effectiveness of the quality management system and to evaluate where continual improvement of the effectiveness of the quality management system can be made. This will include data generated as a result of monitoring and measurement and from other relevant sources.

The analysis of data will provide information including

(a)   customer satisfaction;   ,

(b)   conformity to product or service requirements;

(c)   characteristics and trends of processes, products or services including opportunities for improvement and preventive action;

(d)   . . . . plc suppliers.

8.5   Improvement

8.5.1   Continual improvement

. . . . plc will methodically strive to continually improve the effectiveness of the quality management system through the use of the quality policy, quality objectives, audit results, analysis of data, corrective and preventive actions and management review.

8.5.2   Corrective action

. . . . plc will take action to eliminate the cause of non-conformities in order to prevent recurrence.

Corrective actions will be appropriate to the effects of the non-conformities encountered.

A documented procedure has been established to define requirements for

(a)   reviewing non-conformities (including customer complaints);

(b)   determining the causes of non-conformities;

(c)   evaluating the need for action to ensure that non-conformities do not recur;

(d)   determining and implementing action needed;

(e)   records of the results of action taken, and;

(f)   reviewing corrective action taken.

8.5.3 Preventive action

. . . . plc has documented procedures to try and ensure appropriate actions are taken to eliminate the causes of potential non-conformities in order to prevent their occurrence in the first place.

A documented procedure has been established to define requirements for

(a)   determining potential non-conformities and their causes;

(b)   evaluating the need for action to prevent occurrence of non-conformities;

(c)   determining and implementing action needed;

(d)   records of results of action taken;

(e)   reviewing preventive action taken.

ISO 9001:2000 Quality Registration Step by Step

8.5.3 Preventive action

. . . . plc has documented procedures to try and ensure appropriate actions are

Issue . . . .   Date . . . .

Annex A

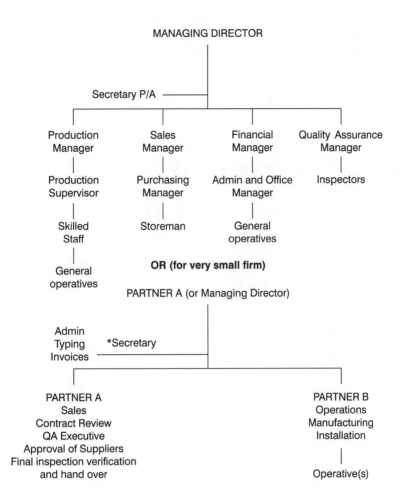

(*Author's Note*: *Secretary may also be Partner A or B if a husband or wife team. Even in a very small firm it is very useful to show FUNCTIONAL reporting routes and identify who is primarily responsible for what.)

ANNEX B
A guide to where the requirements of paragraphs of the Quality Manual have been addressed in the procedures

Procedures	Associated Reference to Para. Q.P.M.
Procedure 1	5.1a, 5.2, 6.1, 6.3, 6.4, 7.1, 7.2.1, 7.2.2, 7.2.3, 7.4.3, 7.5.4 Plus requirements in some certification bodies regulations for 'bought-in-and-supplied-on'
Procedure 2	7.4, 7.5.4, 8.2.4
Procedure 3	7.5, 8.2.3, 8.2.4
Procedure 4	7.5.3, 7.5.4, 8.3
Procedure 5	7.5.1(c), 7.5.1 (d), 7.6
Procedure 6	6.2.1, 6.2.2
Procedure 7	4.1(f), 5.1(c), 5.3(c), 5.4.1, 5.4.2, 5.6, 7.1(a), 7.2.3(c), 7.4.1, 7.5.2, 8.1(c), 8.2.2, 8.3, 8.4, 8.5.1, 8.5.2, 8.5.3
Procedure 8	4.2.1, 4.2.2, 4.2.3, 7.5.4
Procedure 9	4.2.4, 7.5.2, 7.5.3, 8.2.4
Procedure 10	4.1, 5.1(c), 5.3(c), 7.1(a), 8.1(c), 8.5.1
Procedure 11	5.5.2, 5.5.3, 7.2.1, 7.2.3

Annex C
Flowchart showing interaction
between processes

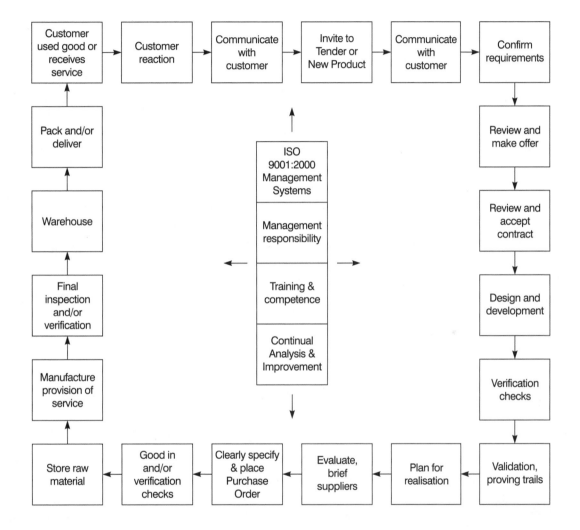

Appendix E

A draft model or workbook to prepare all your required Operating Procedures

Notes

1. Throughout this draft model, at appropriate places, possible alternatives, or additional text is suggested. This may be more applicable to your organisational structure, management style, product or services. These are shown in brackets in the form (**OR** . . . .).

2. Where the text shows . . . . plc, these are appropriate places for you to insert the name of your organisation.

INDEX OR LIST OF .... PLC OPERATING PROCEDURES

OP1   The procedure for review and approval of offers, review and acceptance of contract and subsequent amendments.

OP2   The procedure for approval of suppliers and subcontractors, the placing of Purchase Orders and for receiving goods and services.

OP3   The procedure for controlling .... plc's processes, including in-process and final inspection and testing. These procedures include controls to cover requirements for handling, storage, packing, preservation and delivery.

OP4   The procedure for control of materials, including non-conforming product.

OP5   Procedure for control of measuring and process equipment.

OP6   Procedure for training.

OP7   The procedure for Management Review, Internal Audits and corrective/preventive actions.

OP8   Procedure for Control of Documents/Data.

OP9   Quality, Inspection and Test Records.

OP10   Quality Objectives and continual improvement.

OP11   Internal and external communication.

| List of Procedures: Issue ...... |
| Dated .... / .... / 200 .... |
| Page 1 of 1 |

A draft model or workbook to prepare all your required Operating Procedures

.... plc
OPERATING PROCEDURE NO. 1

The procedure for review and approval of offers, review and acceptance of contracts and any subsequent amendments

Authorisation and amendment record

Issue No.	Date of Issue	Prepared by	Authorised by:
1	1 Jan 20	A.N. Other	Draft for comment
2	1 Feb 20	A.N. Other	Informal Issue
3	1 April 20	A.N. Other	2nd Informal Issue
4	1 July 20	A.N. Other	. . . . Signed: M.Director
5			
6			

Operating Procedure 1:   Issue . . . . . .

Dated . . . . / . . . . / 200 . . . .

Page 1 of 5

.... plc
OPERATING PROCEDURE NO. 1

The procedure for review and approval of offers, review and acceptance of contracts and any subsequent amendments

1. Purpose and scope

The purpose and scope of this procedure is to ensure that:

(a) .... plc, or a member of its staff, does not make an offer, or accept a contract or order, to carry out work (**OR** perform services) that are outside its capability to deliver.

(b) In both cases, before submission of an offer, and also on receipt of a contract, it is carefully reviewed to ensure:

    (i) the CONTRACT is adequately defined

    (ii) the REQUIREMENTS are adequately defined and the appropriate version of any documents, specifications or drawings are held and have also been reviewed;

    (iii) the specification is complete and there are no additional items we think should be included, or necessary for successful performance of the product or service. If we are in any doubt at all, we will always contact the client to ensure the order will fully satisfy their needs.

    (iv) there are no unresolved discrepancies or differences between any offer made by .... plc and the contract now placed;

    (v) that .... plc has the capability to carry out the work (**OR** perform the service) or have ensured that we have approved suppliers or subcontractors for the components, subassemblies or service items outside .... plc's capability;

    (vi) that the materials are held in stock or can be sourced in time from approved suppliers and that this material is one that .... plc considers safe to store and use;

    (vii) that .... plc wishes to carry out the work for the fee offered and under the agreed terms and conditions;

    (viii) that the product or service being provided meets all statutory and legal requirements of the destination country. Also any statutory and legal requirements for packaging, delivery and custom requirements are met.

    (ix) that the proposal/contract is reviewed carefully, to check if there are any material or services being provided by the client as 'free-issue', that need special arrangements for control;

    (x) that the proposal/contract is reviewed, to check if there is a potential requirement of the client demanding that they can inspect any bought-in material or services at source which requirement will be clearly noted in the contract file, to then be addressed as a clause on appropriate .... plc purchase orders.

Operating Procedure 1: Issue ......

Dated .... / .... / 200 ....

Page 2 of 5

2. Responsibility

The only person within . . . . plc who can review and make an offer or accept a contract of a £1000 (**OR** . . . .) or above, is the Managing Director. The Production Manager can carry out a review for the repeat of a sucessful job or below this figure, in the absence of the Managing Director, but must obtain the agreement and the additional signature of the Financial Manager (**OR** . . . .).

3. Implementation

3.1 Making an offer or submission of a tender

To keep track of all invitations, offers and contracts . . . . plc maintain an 'Enquiry and Contract Book'. This record covers two pages of an A4 record book, the headings within this book are shown at Annex A of this procedure.

It is the policy of . . . . plc that all offers will be made in writing by formal letter or by fax. The author of the letter may use Annex B as an additional guide or checklist to ensure all points are covered (**OR** . . . . plc jobs are quite complicated hence an estimating sheet has been developed to plan and cost each potential job see Annex . . . .). Such an offer, letter or fax will be reviewed by the Managing Director, against the points in para 1(b) and authorised by signing and dating the letter and also the 'Enquiry and Contract Book' (**OR** . . . . form). (**OR** unfortunately the type of market that . . . . plc is engaged in dictates that verbal tenders to carry out work must be provided. In these cases the offer will be documented and formalised within . . . . plc by entering the details and signing in the 'Enquiry and Contract Book'.)

When an offer/contract is reviewed by the Production Manager with the Financial Manager (see para. 2), only one will sign the offer document, but both will sign the 'Enquiry and Contract Book'.

3.2 Accepting a Contract

No work will commence or material be ordered until a contract (including, acceptance of a . . . . plc offer, instruction to proceed, etc.) has been formally reviewed and authorised by signing and dating the 'Enquiry and Contract Book'. These are usually done personally by the Managing Director but in certain circumstances can be carried out by others (see para. 2 above).

ALL small value orders below £1000 (**OR** . . . .) will have the Contract Review carried out simply by checking against the items listed in 1(b) and recorded in the 'Enquiry and Contract Book'. For orders of £1000 (**OR** . . . .) or greater, a formal contract review will also be carried out recorded on the form at Annex B of this procedure.

A suitable letter of acknowledgement (or declining letter, if appropriate) is sent to accept a written contract and a confirmation fax is sent to all verbal contracts. (**OR** as standard practice all accepted contracts are acknowledged on form at Annex D with photocopy retained in the file.) (**OR**, our clients do not want to place written orders or accept formal acknowledgements. Whilst . . . . plc find written orders desirable, it is a fact of our market place, that we must accept verbal orders and acknowledgements or our clients will go elsewhere. Hence, orders are formally acknowledged or recorded within . . . . plc via the 'Enquiry and Contract Book'.)

Following a successful contract review a contract file (see Operating Procedure 8, Document Control) is opened and formal work instructions (**OR** job cards) are issued.

Operating Procedure 1: Issue . . . . . .

Dated . . . . / . . . . / 200 . . . .

Page 3 of 5

3.3 'Free-Issue' Material

Ensure that the work instructions carry any requirements to control material supplied as 'free-issue'.

It is also understood that customer property can include interllectual property such as designs, drawings, commercial-in-confidence information, etc. Such customer property will remain 'commercial-in-confidence', it will not be photocopied for supply to others, or used by . . . . plc on our own products.

If we have visitors to . . . . plc, due care will be taken to ensure that the visitors do not have sight of their competitors' specifications or designs.

3.4 Verification by client, of material at source

Ensure that any requirements of the client to visit . . . . plc suppliers is documented in the instructions and included on the appropriate Purchase Orders.

3.5 Amendments to contracts

In order to prevent disputes at a later date the policy of . . . . plc is only to accept authorised (signed) amendments to contract in writing (either letter, memo or fax).

These amendments are also reviewed against the items shown in para 1(b).

If the amendment to contract is acceptable to . . . . plc, the client is advised in writing and also agreement is sought for any additional fee. Before the amendment to contract is actioned, confirmation must be received of any extra payment.

Amendments to contract are reviewed and actioned as follows:

(i) in the 'Enquiry and Contract Book' a neat line is drawn through the existing record of 'year and serial number', only

(ii) in the next space available write the same year and serial number with added amendments reference e.g. 96/123'A' means contract reference 96/123 amendment 'A', 96/456'D' means contract reference 96/456 is on its 4th amendment (i.e. amendment 'D').

Complete the appropriate details across the 'Enquiry and Contract Book', to ensure the amendment is reviewed, accepted and recorded.

Even for large contracts it would not normally be necessary for amendments, to complete Annex B for amendments but it may be completed, if appropriate.

Following completion of the above review the amendments to the formal work instructions are issued, by withdrawing and amending the job card (**OR** . . . .).

Operating Procedure 1:   Issue . . . . . .

Dated . . . . / . . . . / 200 . . . .

Page 4 of 5

---

4. Factored work, or work bought-in and supplied on

It would not normally be the policy of . . . . plc to accept a contract and then subcontract the **WHOLE** of the job to another organisation.

If . . . . plc were faced with this situation we would attempt to find another company with a valid accredited certificate for ISO 9001 covering the scope of work required. If eventually the whole job was placed on a non ISO 9001 registered company, the client would be advised that:

'The . . . . was being manufactured (**OR** supplied) outside the scope of our ISO 9001 registered system'.

(**OR**. In order to complete . . . . plc's product range it is necessary to buy-in complete, some of the smaller models (**OR** . . . . models) and to sell them on as finished items. Every attempt is made to procure these from a firm with a valid accredited registration to ISO 9000 for the products concerned.

These items are strictly controlled to ensure they are the same grade or standard as the . . . . plc products. To accommodate this situation and to make it clear . . . . plc's ISO 9000 scope of registration includes 'stockholding/distribution'. If a client seeks confirmation of the source of these items we will automatically advise they are not manufactured by . . . . plc and whether or not they are from a Quality Assured source.

5. Recording minutes of meetings on site visits

In . . . . plc's type of business some information can only be obtained by site visits and/or given at meetings with the client. It is vital that these be formally recorded, reviewed and actioned where necessary. These visits and meetings (including meetings with the client at . . . . plc) will be recorded on the form at Annex 'C' to this procedure. Alternatively, the client or . . . . plc may provide formal minutes of the meeting. These will be reviewed by the appropriate officer of . . . . plc. The minutes will be signed and dated by the reviewer with notes and actions recorded in manuscript in permanent ink on the minutes, if appropriate.

On take-over of construction sites (**OR** ground maintenance contracts, **OR** . . . .) it may be appropriate to also take photographic or video records.

Operating Procedure 1: Issue . . . . . .

Dated . . . . / . . . . / 200 . . . .

Page 5 of 5

ENQUIRY AND CONTRACT BOOK (Left-Hand Page)

Reference		Enquiry		Quote date	Quote review (sign or initial)	Organisation name, address if appropriate	Client contact (position & Tel. No. if appropriate)
Year	Serial No.	Date	Written or verbal				

Annex A Oper. Proc. 1:   Issue . . . . . .

Dated . . . . / . . . . / 200 . . . .

Page 1 of 2

ENQUIRY AND CONTRACT BOOK (Right-Hand Page)

1. Detail & Scope of Work 2. Delivery Date 3. Agreed Price 4. Amendment	Contract review/ acceptance date	Contract review (sign or initial)	Contract file opened (sign)

Annex A Oper. Proc. 1: Issue . . . . . .

Dated . . . . / . . . . / 200 . . . .

Page 2 of 2

LARGE VALUE OR RISK/CONTRACT REVIEW AND ACCEPTANCE

Year . . . .   Serial No. . . . .

Organisation: . . . .   Approx. Value: . . . .

Enter YES/NO, Details

1. Letter of unconditional acceptance (or contract)
   reviewed
   OR
   Verbal contract (confirmation letter/fax sent?)   . . . .

2. Is this contract different from any offer made by
   . . . . plc?   . . . .

3. Methods of payment specified   . . . .

   Are they acceptable to . . . . plc?   . . . .

4. Any special specified terms e.g. firm price, variable,
   COD, penalty clauses, liquidation damages, adverse
   retention, insurance liability, inspection visits to our
   suppliers, etc.   . . . .

5. Is the contract on . . . . plc terms of business, or
   the clients?   . . . .

   Is this acceptable to . . . . plc?   . . . .

6. Is the required scope of work clear and
   unambiguous?   . . . .

7. Are all specifications/drawings and other details
   available and can their requirements be met?   . . . .

8. Are there any small print clauses? Have these been
   reviewed and are they acceptable?   . . . .

9. Confirm that . . . . plc have the resources, skills,
   equipment and ability to supply.   . . . .

10. Where we need the input or assistance of outside
    resources have we arrangements with approved
    subcontractors or suppliers?   . . . .

    Or can we obtain the same from a guaranteed
    source?   . . . .

Annex B Oper. Proc. 1: Issue . . . . . .

Dated . . . . / . . . . / 200 . . . .

Page 1 of 2

11. Statutory requirements, Health & Safety issues,
    environmental or 'CE' implications   . . . .

12. Any 'free-issue' material supplies and controls
    required?   . . . .

13. Are site visits required and reports available?   . . . .

14. Can . . . . plc meet the delivery schedule?   . . . .

    Design, development & drawings lead time   . . . . weeks

    Order raw material   . . . . weeks

    Obtain buildings/machinery   . . . . weeks

    Any special work/environmental requirements . . . . weeks

    Manufacture   . . . . weeks

    Test & inspect   . . . . weeks

    Packaging or delivery requirements   . . . . weeks

    Other items

    . . . . weeks

    . . . . weeks

    . . . . weeks

    Estimated earliest delivery   . . . . weeks

15. Other items of concern:

    . . . .

    . . . .

16. Is there a requirement for a formal 'Quality Plan'?   . . . .

    Contract Accepted . . . .
                         Signed   Dated

17. (i)   Send letter of acknowledgement (to written contract)
    (ii)   Send Fax acknowledgement (to verbal order)
    (iii) Enter details and sign 'Enquiry & Contract Register Book'
    (iv)   Open Contract File
           File this review in Contract File
           Raise Work Instructions (**OR** job card) (**OR** . . . . .)

Annex B Oper. Proc. 1: Issue . . . . . .

Dated . . . . / . . . . / 200 . . . .

Page 2 of 2

.... plc
SITE VISIT/MEETING REPORT

Circulate to:
Contract/Enquiry file
Client (if appropriate)
Managing Director

Year . . . .   Serial No. . . . .

Organisation   . . . .

Outline description of project/contract   . . . .

Visit to/meeting at   . . . .

Person(s) representing . . . . plc   . . . .

Person(s) representing client organisation   . . . .

Others present   . . . .

Date (and time where appropriate)   . . . .

Record of observations/discussions:

Additional pages **YES/NO**

Actions arising: **Actioned by:**

Recorded by . . . . date . . . .
Signed

Annex C Oper. Proc. 1: Issue . . . . . .

Dated . . . . / . . . . / 200 . . . .

Page 1 of 1

ACKNOWLEDGEMENT OF ORDER

.... plc

....

....

Tel: XXXX XXXXXXX

Fax: XXXX XXXXXX4

To: Mr/Mrs/Dr ....

....

....

Postcode: ....

We acknowledge with thanks receipt of your order Ref: .... for

....

....

....

....

At: .... Site

Estimated/Contractual start date: ....

In accordance with the terms and conditions set out in ....

dated ....

In any future correspondence with .... plc regarding this contract,

please quote our ref no: ....

The Project Manager at .... plc for this contract is ....

....
Signed   Dated

Annex D Oper. Proc. 1:   Issue ......

Dated ..../..../ 200....

Page 1 of 1

.... plc
OPERATING PROCEDURE NO. 2

The procedure for approval of suppliers and subcontractors, the placing of purchase orders, and for receiving goods and services

Authorisation and amendment record

Issue No.	Date of Issue	Prepared by	Authorised by:
1	6 Feb 20	A.N. Other	Draft for comment
2	12 Feb 20	A.N. Other	2nd Draft for comment
3	20 Feb 20	A.N. Other/J. Smith	Informal Issue
4	1 March 20	J. Smith	. . . . Signed: M. Director
5			
6			

Operating Procedure 2: Issue . . . . . .

Dated . . . . / . . . . / 200 . . . .

Page 1 of 9

.... plc
OPERATING PROCEDURE NO. 2

The procedure for approval of suppliers and subcontractors, the placing of purchase orders, and for receiving goods and services

1. Purpose and scope

The purpose and scope of this procedure is to ensure that:

(a) the staff of .... plc do not buy shoddy or inferior goods or services from the cheapest supplier. The goods or services bought must meet .... plc's and its customer's requirements. The policy of .... plc is to purchase from sources that consistently deliver on time, goods and services to the required specification at a fair and reasonable price.

(b) all purchasing by .... plc is carried out in a professional manner with adequate records kept. For the purpose of ISO 9001, this procedure need only apply to items affecting the quality of our product or service. However, to give a common approach all purchasing within .... plc will follow these procedures.

2. Responsibility

2.1

Any member of .... plc may require and request raw material, tools, goods or services to be purchased (**OR** there may also be a purchase demand from Stock Control reaching a re-order quantity for the stores stock.) (**OR** this request is made on a Purchase Requisition Form, see Annex ....)

Following such a request a formal purchase order may be written out by the storeman or any manager. In practice the majority are raised by the Office Manager, (**OR** ....). (**OR** may only be placed by the Purchasing Manager or a Director.)

2.2

Before an order can be placed it must be reviewed and then signed for authorisation to place the order outside .... plc.

It should be noted that this procedure outlines the only approved method of raising purchases on behalf of .... plc.

Operating Procedure 2: Issue ......

Dated ..../..../ 200....

Page 2 of 9

Up to and including a value of £100 the purchase order can be reviewed and authorised by a manager (**OR** by . . . .).

All orders above £100 must be authorised by a Director (**OR** . . . .).

(**OR** for larger organisations) the levels of authority for review and signature are:

	Up to £50	£51 to £250	£251 to £1000	£1001 to £5000	£5001 & above
Items in aid of current job, previously fully specified and detailed on drawings, etc.	PM PS	PM PS	PM PS/FM	MD PM/FM	MD
General raw materials in aid of production	PM PS St/PS St/OM	MD PM PS/FM	MD PM	MD	MD
Hand tools, consumables, clothing, janitorial items, etc.	PM PS OM St/PS St/OM	PM OM	PM/FM OM/FM	MD	MD
Printed material & stationery	PM OM	OM	MD OM/FM	MD	MD
Capital plant & machine tools	PM	PM	PM/FM	MD	MD

MD	=	Managing Director
PM	=	Production Manager
PS	=	Production Supervisor
FM	=	Finance Manager
OM	=	Office Manager
St	=	Storeman
PM/FM	=	Signature by both on Purchase Order, by exception, to cover situation where normal responsible officer is off site.

Whilst . . . . plc have a Purchasing Manager his responsibility is to procure the goods. The Purchasing Manager cannot review and authorise Purchase Orders.

2.3

Only the Quality Assurance Manager (**OR** Managing Director, **OR** . . . .) can review and authorise additional suppliers onto the Approved Suppliers List.

2.4

Goods receiving inspection will be carried out by the Quality Manager (**OR** Site Foreman, **OR** . . . .) and in his/her absence by . . . . (**OR** by the person who raised the Purchase Request).

3. Implementation

3.1 Approval of . . . . plc suppliers

Items, material or services that affect the quality of the product or services that are provided by . . . . plc shall be purchased from suppliers or subcontractors that are on the formal 'Approved Suppliers List'.

This list is given an issue number and date of issue and is authorised by the Managing Director (**OR** QA Manager **OR** . . . .) to authorise its issue. To control its issue the list is attached as Annex A to this procedure (**OR** to control its issue and to ensure that the same copy is used throughout the organisation it is considered a controlled document and controlled as per Operating Procedure 8).

3.2

To gain entry onto the . . . . plc 'Approved Suppliers List' the firm, organisation, consultant, certification body, or sole trader will be formally evaluated and recorded on form Annex 'B' to demonstrate at least one of the following:

   (i)   They are registered or certified by a recognised UKAS accredited independent third party body to ISO 9001 with a scope of registration equivalent to the products/ services that . . . . plc require. Alternatively, second party approval (i.e. approved by a recognised main contractor e.g. British Rail, British Nuclear Fuels, etc.) may be judged to be acceptable for appropriate products or services.

   (ii) There exists records that can demonstrate that the subcontractor or supplier has historic evidence of providing quality goods or services, to . . . . plc consistently over at least 5 orders (**OR** . . . . orders) during the last 6 months (**OR** . . . . months) prior to adding them to the list.

   (iii) The supplier is an official distributor or approved agent for the material, parts or service of a proprietary product.

   (iv) That as a potentially new supplier to . . . . plc the organisation has been carefully evaluated for suitability and confirmation of ability to supply consistently to . . . . plc specifications. This may be by trial order or audit visit; the evaluation will be documented on the Supplier Evaluation Record (Annex 'B').

   (v)   Special qualification. To be detailed by memo by Quality Assurance Manager (**OR** . . . .), which will be signed and dated and stapled to the Supplier Evaluation Record.

Record sheets are kept for each supplier/subcontractor to demonstrate the above evaluation, see Annex B to this procedure. The stages on Annex B are self-explanatory and go through the alternative methods of approval (i)–(v) above. Where a questionnaire is required this is shown at Annex C.

Operating Procedure 2:   Issue . . . . . .

Dated . . . . / . . . . / 200 . . . .

Page 4 of 9

3.3

If a audit or confidential visit is made to a potential supplier it is recorded on the form shown at Annex C of Operating Procedure 1.

The items to check and report on during the visit should include the following, where appropriate:

  (i)   Quality of management and systems;
  (ii)   Systems documentation;
  (iii)   Quality, training and ability of staff and operatives;
  (iv)   Process control and capability of their equipment;
  (v)   Storage facilities;
  (vi)   Typical products/service and check against specification;
  (vii)   Reject rates;
  (viii) Check against the clauses of ISO 9001 where they apply;
  (ix)   Size, location, condition of equipment and buildings;
  (x)   Stockholding of materials and spares etc;
  (xi)   Financial viability.

(**OR** for critical components materials, the quality of the supply of . . . . and . . . . .are absolutely critical to . . . . plc, therefore all potential and existing suppliers are/were visited and audited before an order is placed. They are audited at least every . . . . . months with formal reports raised, to confirm their continued suitability.)

4. Procedure of placing a purchase order

4.1

It is the policy of . . . . plc that all purchase orders be written, reviewed and authorised before the order is actually placed. (**OR** if the value of the purchase is above £ . . . . . a competitive quote will be obtained), (**OR** the Managing Director's advice sought.)

4.2

Occasionally and by exception a telephone order has to be placed to meet an urgent or emergency need. Such an order will be written down as soon as possible and endorsed 'Confirmation of telephone order by our Mr./Mrs. . . . . of the . . . . / . . . . /200 . . . . '.

4.3

As . . . . plc does not place many orders and they are generally of brief description it has been found that a standard A5 stationery triplicate book with pre-printed serial numbered pages is entirely satisfactory for our needs. (**OR**, the order will be written on a formal . . . . plc Purchase Order Form, see Annex D.)

(Author's Note 1. The Purchase Order form at annex 'D', shows the typical information and statements required. You should reduce fonts as required to obtain maximum space for you to write or type the appropriate details. Item (#), if included, ensures clause 7.4.3 is automatically covered.)

(Author's Note 2. It may well be advisable in your business to have formal legally drafted 'Terms and Conditions' printed on the front or reverse of your Purchase Order, to ensure goods/materials are always supplied to you on your own Terms and Conditions. In some industries the Terms and Conditions of suppliers and also their carriers are very restrictive; especially to return or reject goods.)

Operating Procedure 2:   Issue . . . . . .

Dated . . . . / . . . . / 200 . . . .

Page 5 of 9

The Purchase Orders will have a unique serial number. The order must be clearly legible and specify the supplier and the details of what are required. The entry must be sufficient to unmistakably identify the supplier and also the description must uniquely specify, without any doubt, precisely what is to be supplied. Use can be made of suppliers brand names or catalogue numbers if appropriate. If a full engineering, technical or chemical specification is required, this will be written and detailed on the order, or referenced with copy attached.

Items to be identified on a purchase order may include:

(a) Purchase Order Number;
(b) Supplier name;
(c) Date of Order;
(d) Supplier address;
(e) Goods, material or service required;
(f) Any appropriate specification or part or drawing numbers;
(g) Level, grade, type, etc.;
(h) Agreed fee or price (recommend clearly state whether with or without VAT);
(i) Delivery date and address: mode of transport where appropriate;
(j) Quantity;
(k) Inspection, certification or special requirements;
(l) .... plc project description or job number;
(m) Markings, identification, or special pack, if required.

(**OR** a copy of this page(s) with the above list will be placed adjacent to .... in a plastic sleeve, to provide a checklist as purchase orders are made out. This will be personally controlled by the Quality Manager (**OR** ....) to ensure it is the correct issue.)

If a .... plc customer has made a specification on raw material or components, this becomes the minimum required of any Purchase Order. If special QA arrangements, test or verified product documentation are required, they must be specified on the Purchase Order.

If it is considered desirable that a representative of .... plc inspect the supplier's processes, systems or the items/material before despatch this will be specified on the Purchase Order. Also if it is considered possible or likely that the eventual .... plc customer for these items/material will wish to visit our supplier this also will be specified as a condition of the Purchase Order.

4.4
One of the persons designated in para 2 will review the order to ensure:

(i) the proposed supplier is on the Approved List;
(ii) the detail given appears to be correct and adequate;
(iii) the appropriate funds are available for this project/purchase order and it is agreed that this may be spent.

If they are satisfied they will sign and date the Purchase Order in the Triplicate Book (**OR** Purchase Order Form). The top copy will go to the supplier (it may be photocopied and faxed, if required), the bottom copy will stay in the Purchase Order Book (**OR** outstanding

Operating Procedure 2: Issue ......

Dated ..../..../ 200....

Page 6 of 9

Purchase Order File). The middle copy will be sent to the goods inward area so they can check the items delivered and the suppliers advice and/or delivery note against the original Purchase Order (**OR** some of . . . . plc's suppliers, as a condition of their trading, will not accept written orders. In these cases the top copy will remain in the Triplicate Book endorsed 'Verbal Purchase Order'.)

(*Author's Note: verbal orders are not a desirable situation but in some cases they are a fact of life; typical examples are foundries buying metal from scrap merchants, purchases of spares by telephone from standard catalogue for small radio and electronic components, purchase of fresh vegetables or meat from traders, etc.*)

4.5

Purchases from unapproved suppliers. Occasionally and by exception it is sometimes essential to buy from an unapproved supplier. There may be several reasons, e.g. they are the sole supplier, emergency breakdown and they are on location, one-off purchase of uncommon item, etc. If this happened **all** copies of the Purchase Order (including the copy for the supplier) will be endorsed 'unapproved supplier, items to be inspected carefully on receipt by . . . .', or other appropriate checks or verification.

4.6

Occasionally our customer will specify a supplier (or model/material) who is not approved by . . . . plc. This will be treated as a condition of the contract placed on . . . . plc, such orders will not be endorsed, refer to unapproved suppliers in previous paragraph. If it is considered or subsequently found that the material or items specified by the customer are not available, are illegal, will not perform, or give unacceptable health and safety risks, etc., the client will be requested in writing to approve an alternative (**OR** . . . . plc will raise a formal concession).

4.7

Amendments to Purchase Orders will **always** be reviewed and if possible be in writing either by memo or fax. A copy of the amendment will be sent to the supplier, with copy attached to the office copy of the Purchase Order and an additional one for the goods inward copy of the Purchase Order.

4.8

(**OR**, site staff are authorised to make exceptional purchases of standard proprietary items or materials up to the value of £50, if required to keep the job going, or to time, to satisfy the customer. This will be reviewed by the Contract Manager and the monies refunded. If the items or materials are considered unsatisfactory by the Contract Manager, they will be replaced at no additional cost to the client.)

5. Procedure for goods inward inspection

5.1

When the delivery has been inspected, to check both the delivery note and the goods provided satisfy the Purchase Order requirements, the goods inward copy of the Purchase Order (**OR** the delivery note) is signed (**OR** stamped), and dated and passed to . . . . to

Operating Procedure 2: Issue . . . . . .

Dated . . . . / . . . . / 200 . . . .

Page 7 of 9

clear the invoice. These copies are retained in a lever arch file to maintain a simple record of Goods Inward Inspection. (**OR** the record of Goods Inward Inspection of the quality-critical items is maintained on the cards shown at Annex E.) (**OR** due to the quantities and variety of items being delivered it has been found useful to maintain a Goods Inward Record Book. This takes two pages of A4 with the headings shown at Annex F.)

The member of staff requiring the goods is informed by phone that the ordered items have been delivered and may be collected (**OR** will be delivered) (**OR** will be booked into the stores).

5.2
If defective items are found they will be dealt with as Operating Procedure 4 'Control of Material' and Operating Procedure 7 'Corrective & Preventive Actions'.

5.3
As the quantities delivered are small they will be inspected 100% (**OR** as some deliveries are large a statistically based sampling plan has been developed using ISO 2859–1:1999.

5.4
As . . . . plc is a small firm with relatively few suppliers, a defective delivery or a poor supplier has a serious effect on . . . . plc production. Hence a poor supplier is automatically identified, therefore no formal vendor rating or scoring system is required.

5.5
All items arriving at . . . . plc will be unboxed and checked for damage and visual appearance. Due note will be taken of any special requirement on the Purchase Order or printed on the product or manufacturer's label. If a dimensional check or laboratory test is required on receipt this will be specified by the person raising the Purchase Order and will be on the Goods Inward copy of the Purchase Order (**OR** by memo that is signed and dated), (**OR** items designated 'Quality Critical' items will be identified on a list authorised by the Quality Manager. For these items a Goods Inward Record Card see Annex E, is completed. This card specifies and records the tests carried out on each delivery and the sample tested, etc). (**OR** items at Goods Inward will be verified and recorded as per the 'Standing Work Instruction' raised, see Annex . . . . , [see Annexes of Operating Procedure 3 for typical examples of work instructions]).

(**OR** . . . . , and . . . . are delivered to . . . . plc in a designed protective pack. On balance it has been found better to simply check the box for damage and quantity and the specification on the label/ advice notes and leave the items within their protective pack.)

5.6 Urgent release of material
No incoming items or material should be used until cleared goods inward inspection. See Operating Procedure 4 para 3.4 for emergency or urgent use.

Operating Procedure 2: Issue . . . . . .

Dated . . . . / . . . . / 200 . . . .

Page 8 of 9

5.7 Delivery of 'free-issue' material or customer property

This will have limited visual inspection. Free Issue material and items WILL be recorded in the Goods Inward Records to demonstrate acceptance and control. They will be labelled as per Operating Procedure 4, or any special authorised instructions. The checks will be limited to:

(i) ensure adequately labelled or identified to prevent misuse or disposal without customer's permission;

(ii) check quantities and advise any shortages;

(iii) check if items are correct to type or description;

(iv) if to be stored where they are not seen regularly, a periodic inspection for deterioration, (which will be recorded).

It is also understood that customer property can include intellectual property such as designs, drawings, commercial-in-confidence information, etc. Such customer property will remain 'commercial-in-confidence', it will not be photocopied for supply to others, or used by . . . . plc on our own products.

If we have visitors to . . . . plc, due care will be taken to ensure that the visitors do not have sight of their competitors' specifications or designs.

Operating Procedure 2:   Issue . . . . . .

Dated . . . . / . . . . / 200 . . . .

Page 9 of 9

A draft model or workbook to prepare all your required Operating Procedures

APPROVED SUPPLIERS LIST

Firm	Approved to supply	Address	Postcode	Tel/fax	Contact

Annex A Oper. Proc. 2:   Issue . . . . . .

Dated . . . . / . . . . / 200 . . . .

Page 1 of 1

165

EVALUATION RECORD OF SUPPLIER TO .... plc (Sheet 1)

Firm: . . . .   Tel: . . . .

Address: . . . .   Fax: . . . .

. . . .   Contact: . . . .

Postcode: . . . .   Position: . . . .

Products/services to be supplied to . . . . plc

Registered with . . . . Cert No. . . . .
photocopy attached confirming scope of registration covers above products/services

The above firm, whilst not registered, has supplied the above services to a satisfactory standard over at least 5 orders
(**OR** . . . . . . orders)

Most recent:
P/O number . . . . dated . . . .

delivery date . . . . . . (**and/or** RIC No. . . . .)

P/O Number . . . . dated . . . .

delivery date . . . . RIC . . . .

P/O Number . . . . dated . . . .

delivery date . . . . RIC . . . .

P/O Number . . . . dated . . . .

delivery date . . . . RIC . . . .

P/O Number . . . . dated . . . .

delivery date . . . . RIC . . . .

Any rejects/comments/remarks

. . . .

. . . .

. . . .

. . . .

. . . .

. . . .

. . . .

. . . .

. . . .

. . . .

Annex B Oper. Proc. 2:   Issue . . . . . .

Dated . . . . / . . . . / 200 . . . .

Page 1 of 2

EVALUATION RECORD OF SUPPLIER TO .... plc (Sheet 2)

The above firm does not appear to be ISO 9000 registered and has no in-house record of consistent good quality supply

Questionnaire sent: date ....   Returned ....

Evaluate and consider criticality of goods/services. Is reply satisfactory **ON ITS OWN** to give confidence to add to .... plc approved suppliers list

YES/NO ....

Audit visit arranged and carried out on ..... *date(s)* see visit report(s) attached.

From evaluation of reports add to approved suppliers list YES/NO ....

Alternatively or in addition to visit:

.... plc placed ...... (*number*) of trial orders on the above firm. The Purchase Orders were amended to specifically show 'This is a trial order to be inspected by .... on delivery' (or before delivery or other special arrangements).

The results are shown on the bottom of page 1 of this evaluation record, with any other appropriate records or details attached.

From evaluation of delivery performance add to Approved Suppliers List
YES/NO ....

Listed official distributor/agent of ....

Add to .... plc Approved Suppliers List

Sign .... Date ....

Removed/deleted from Approved Suppliers List

Reason/Remarks (if appropriate)

Sign .... Date ....

Annex B Oper. Proc. 2:   Issue ......

Dated ..../..../ 200....

Page 2 of 2

. . . . plc
Tinatown Industrial Estate
Tinatown
Blankshire
BL12 3PM

Date: . . . .
Tel: . . . .
Fax: . . . .

Mr. . . . .
. . . . Ltd.
. . . .
. . . .

Dear Mr./Mrs . . . .

Questionnaire to supply to . . . . plc

The directors of the above firm are committed to obtaining quality in our products and services. We also are seeking (**OR** intend to maintain) ISO 9000 registration. Obviously a key element of this policy is to ensure that our purchased items, materials and services are to a similar standard.

To this end and to ensure future trading, please can you complete the form attached and return.

If you have any queries or need any assistance please contact . . . .

Yours sincerely,

Managing Director
. . . . plc

Annex C Oper. Proc. 2: Issue . . . . . .
Dated . . . . / . . . . / 200 . . . .
Page 1 of 6

QUESTIONNAIRE TO AID EVALUATION OF SUPPLIER TO . . . . PLC

Part A to be completed by . . . . plc

Firm: . . . .

Address: . . . .

. . . .

Postcode: . . . .

Products/services proposed to be purchased by . . . . plc

. . . .

. . . .

Part B to be completed by potential supplier (please delete items not relevant)

Name and address if different from above:

. . . .   Tel. No: . . . .

. . . .   Fax No: . . . .

Preferred Contacts:

1. . . . .   2. . . . .   3. . . . .

Positions: . . . .

We confirm that we can supply the item(s) at Part A

Signed . . . .   Date . . . .

We can supply the above with the following conditions/amendments:

Signed . . . .   Date . . . .

Please note we can also provide:

Approx annual turnover . . . . Number of full time staff . . . .

Number of part time staff . . . .

Annex C Oper. Proc. 2:   Issue . . . . . .

Dated . . . . / . . . . / 200 . . . .

Page 2 of 6

QUESTIONNAIRE TO AID EVALUATION OF SUPPLIER TO .... PLC

Part C to be completed if appropriate by potential supplier

We are ISO 9000 registered by a UKAS accredited body for the products/services above by
.... Certificate No. .... please find photocopy attached.

Signed .... Dated ....

Note: If you can answer Part 'C' positively, please cross through the remainder of the form
and return to .... plc. Please ensure that the copy of certificate provided gives the
scope of registration. A presentation type certificate without the scope is considered to be
meaningless. Similarly a certificate from an unaccredited body is totally unacceptable.

Part D to be completed if appropriate by potential supplier

(i)   The firm is in the process of implementing Quality Assurance Systems based on ISO
      9000 standards and will be seeking assessment and certification in the month of
      .... 20 ... Please find attached photocopy letter from .... confirming
      that we have made such an application for independent third party UKAS accredited
      assessment/certification.

      Tick,   YES ......   Not applicable ......

(ii) The firm has no immediate plans at this time to apply for independent third party
      assessment and certification of our management systems to provide independent
      confirmation of appropriate and effective Quality Assurance arrangements.

      Tick,   No Plans ......   May in future ......

(iii) Where applicable, what controls (systems, procedures, instructions, equipment, etc.)
      do you have for the following areas:

      1. How do you manage your organisation and define the areas of responsibility?

      ....
      ....
      ....

      2. Do you have a defined person, who irrespective of other duties, has
         responsibility for Quality Assurance?

      ....
      ....
      ....

Annex C Oper. Proc. 2:   Issue ......

Dated .... / .... / 200 ....

Page 3 of 6

3. How do you review your progress and performance including Quality Assurance performance?

. . . .

. . . .

. . . .

4. Do you have processes for continual improvement?

. . . .

. . . .

. . . .

5. How do you prepare and plan to ensure you have sufficient resources of staff/equipment and expertise including Quality Assurance resources?

. . . .

. . . .

. . . .

6. How do you ensure that you only accept contracts suitable for your company and that you have enough information to proceed?

. . . .

. . . .

. . . .

7. What processes have you to ensure you comunicate with your customer to ensure they are satisfied?

. . . .

. . . .

. . . .

8. If applicable, what procedures do you employ to ensure you control your design process?

. . . .

. . . .

. . . .

9. How do you control Quality Assurance and contract documents including drawings and specifications to ensure all the copies in use are the same issue?

. . . .

. . . .

. . . .

Annex C Oper. Proc. 2:   Issue . . . . . .

Dated . . . . / . . . . / 200 . . . .

Page 4 of 6

10. How do you control your suppliers and subcontractors?

   . . . .

   . . . .

   . . . .

11. What controls do you have for identifying products (including material supplied by . . . . plc in aid of a contract). Also, how do you identify what is good from the doubtful and segregate non-conforming product?

   . . . .

   . . . .

   . . . .

12. How do you control your processes to ensure goods/services are to specification?

   . . . .

   . . . .

   . . . .

13. What provision do you have for inspection and testing?

   . . . .

   . . . .

   . . . .

14. If you are using equipment or instruments to check the product/services how do you ensure that the instrument readings/results are correct?

   . . . .

   . . . .

   . . . .

15. What controls do you have for short-term corrective action?

   . . . .

   . . . .

   . . . .

   Long-term preventive actions

   . . . .

   . . . .

   . . . .

Annex C Oper. Proc. 2:   Issue . . . . . .

Dated . . . . / . . . . / 200 . . . .

Page 5 of 6

16. What provisions do you have for handling, storage, packing, preservation and delivery?

. . . .

. . . .

. . . .

17. What records are kept to demonstrate process control, inspection, etc.?

. . . .

. . . .

. . . .

18. What provisions do you make to carry out audits of your Quality Assurance systems and procedures to ensure these are effective?

. . . .

. . . .

. . . .

19. Have you training records to demonstrate that your staff are competent, adequately trained and/or qualified?

. . . .

. . . .

. . . .

Other comments:

. . . .

. . . .

. . . .

Are you prepared to allow, if considered necessary, a representative of . . . . plc to make arrangements to visit your site to confirm your suitability to supply?

YES/NO . . . .

Signed . . . .   Position . . . .

Date . . . .

Annex C Oper. Proc. 2:   Issue . . . . . .
Dated . . . . / . . . . / 200 . . . .
Page 6 of 6

ALTERNATIVE ONE PAGE QUESTIONNAIRE

Company Name: . . . .

Product: . . . .

1. Do you have a Documented Quality System? YES/NO

2. Is the system approved by an accredited Certification body? YES/NO
   (If YES, attach certificate which details the scope of approval,
   ignore the remaining questions and sign the form and return it)

3. Do you carry out and record Contract Reviews and have processes
   for communicating with your customers? YES/NO

4. Do you have a system for controlling documents such as drawings
   and specifications? YES/NO

5. Do you have a formally evaluated and controlled approved list of
   suppliers and purchasing system? YES/NO

6. Do you maintain product identity? YES/NO

7. Do you maintain product traceability? YES/NO

8. Do you have formal inspection, with appropriate records? YES/NO

9. Do you operate a formal calibration system? YES/NO

10. Do you formally control and review non-conformances? YES/NO

11. Do you have a formal customer complaint register? YES/NO

12. Do you have a formal system of Internal Audits? YES/NO

13. Do you have a formal system for corrective/preventive actions
    and setting quality objectives and continual improvement? YES/NO

14. Do you have a formal system of reviewing and satisfying training needs? YES/NO

15. How long do you retain inspection records? . . . . . years

16. Total number of employees . . . .

    Total number in production . . . . , number in QA/Inspection . . . .

. . . .
      Signed   Position   Date

PURCHASE ORDER

Motif or logo of
. . . . plc

From: . . . . plc
Tinatown Industrial Estate
Tinatown   Order No: . . . .
Blankshire   Date: . . . .
BL12 3PM   Job No: . . . .
Tel: . . . .   Title: . . . .
Fax: . . . .   Date Required by: . . . .
VAT No. . . . .

To:

Please supply the goods/material/services detailed below in accordance with: (**OR** our standard terms and conditions overleaf and)

Your standard catalogue, your agreed fee sheet, your written quotation of . . . ./. . . ./199 . . . , your verbal quotation to our Mr/Mrs . . . . of . . . ./. . . ./199 . . . or . . . .

Item	Details/Specification/Part No. (incl. inspection, certification or special requirements)	Qty	Unit	Price per unit	Tax

* Delivery to address above, or . . . .

* Please supply to following Certificates . . . .

* Please quote above order reference in correspondence or invoice. All invoices paid within 12 working days of the end of the month that the invoice is presented.

* . . . . plc reserve the right to inspect the items above during manufacture or at source, prior to delivery. Such inspection may be by . . . . plc directly, our approved agents or a representative of our customers (#)

Reviewed and approved . . . . (sign)   date . . . .

(* Delete as appropriate)

Annex D Oper. Proc. 2:   Issue . . . . . .

Dated . . . ./. . . ./ 200 . . . .

Page 1 of 1

.... plc

GOODS INWARD RECORD CARD
DESIGNATED QUALITY CRITICAL ITEMS

Description/Specification ....

....

....

Part No: .... Stores No: ....

Drawing No: .... Issue of drawing (**or** date) ....

Supplier: .... Tel: ....

.... Fax: ....

Contacts Name: ....

Position: ....

Special Tests or gauges required:

....

....

Date	Del. note	RIC No.	Qty del.	Serial No. gauge	Comments, No. accepted/rejected, special features, documentation reference, etc.

Annex E Oper. Proc. 2: Issue ......

Dated .... / .... / 200 ....

Page 1 of 2

CONTINUATION SHEET
GOODS INWARD RECORD CARD

Supplier:. . . . .

Part/Stores/Drawing No. . . . .

Date	Del. note	RIC No.	Qty del.	Serial No. gauge	Comments, No. accepted/rejected, special features, documentation reference, etc.

Annex E Oper. Proc. 2: Issue . . . . . .

Dated . . . . / . . . . / 200 . . . .

Page 2 of 2

GOODS INWARD RECORD BOOK

(Left-Hand Page)

Date of delivery	Carrier	Advice Note Ref.	Part No.	Description	Qty

Annex F Oper. Proc. 2: Issue . . . . . .

Dated . . . . / . . . . / 200 . . . .

Page 1 of 2

GOODS INWARD RECORD BOOK

(Right-Hand Page)

Sample size taken	No. accepted	No. rejected	Supplier	Comments, Concessions, Certs. of Conformity Material Cert., Test Certificate references, etc.

Annex F Oper. Proc. 2:   Issue . . . . . .

Dated . . . . / . . . . / 200 . . . .

Page 2 of 2

. . . . plc
OPERATING PROCEDURE NO. 3

The procedure for controlling . . . . plc's processes, including in-process and final inspection and testing. These procedures include controls to cover requirements for handling, storage, packing, preservation and delivery

Authorisation and Amendment Record

Issue No.	Date of issue	Prepared by	Authorised by:
1	1 Feb 20	A.N. Other	Initial Draft
2	12 Feb 20	A.N. Other	2nd Draft
3	2 March 20	A.N. Other/J. Smith	Informal Issue
4	5 April 20	J. Smith	. . . .
			Signed: M. Director
5			
6			

Operating Procedure 3:   Issue . . . . . .

Dated . . . . / . . . . / 200 . . . .

Page 1 of 10

.... plc
OPERATING PROCEDURE NO. 3

The procedure for controlling .... plc's processes, including in-process and final inspection and testing. These procedures include controls to cover requirements for handling, storage, packing, preservation and delivery

1. Purpose and scope

The purpose and scope of this procedure is to ensure that:

(a) all the processes of .... plc which affect the quality of the delivered product (**OR** service) are carried out under controlled conditions to ensure that the output consistently meets the required specification.

(b) as required, in-process and final inspection are specified and controlled.

(c) as required, the procedure provides controls to ensure products are handled, packed, stored and delivered to specification or in a manner that prevents deterioration and/ or danger.

2. Responsibility

This procedure has been authorised by the Managing Director. However, the Production Manager (**OR** Contracts, **OR** Operations **OR** ....) is responsible for ensuring its continued suitability and to advise the Quality Manager of any changes to documentation that are required.

The Production Manager and Production Supervisor (**OR** ....) must ensure these procedures are implemented fully.

It is the responsibility of all staff and operatives to comply fully with the requirements of this procedure.

(Author's note: there now follows six different models for documenting Process Control Procedures:

3   Model for Production/Manufacturing with one basic process with several sizes/models of the same product.

3A   Model for Production/Manufacturing with a variety of completely different processes and inspections or test.

3B   Model for Metal Machining, Fabrication or Vehicle Body Shop with jobs of small batch size or one-offs.

3C   Model for Site Contractor, Landscaper, Electrical Installation, Construction, Plastering, Painting, Office Cleaning, etc.

Operating Procedure 3:   Issue ......

Dated .... / .... / 200 ....

Page 2 of 10

3D Model for well regulated industry, profession or service, including approved agents for Ford, Konica, Apricot, etc., etc.

3E Model for organisations who already have adequately and formally documented process control procedures.

Review these and choose the one nearest to how you manage your business and then document your process controls with adequate inspection and verification stages. Do scan through the others and pick and mix to achieve the controls you deem to be necessary.

Do not try and second guess the assessors and put in what you they may want: you do it how you want; they have to try and show that this is not adequate in order to raise a non-conformance.)

3. Implementation (model for Production/Manufacturing shop with one basic process with several models/sizes of similar products)

.... plc make a standard range of products. This goes through the following planned stages and sequence of process events and inspection checks, with the individual variations as detailed in the Manufacturing and Testing specification for .... product (**OR** product ...., product .... product .... etc.):

3.1 The machinist books the raw material out of stores on form .... (**OR** recorded in issue book), for each production order. The material is kitted into sets .... etc. as detailed in the specification for that model.

3.2 Preliminary machining takes place on .... using jigs reference .... (**OR** to drawing numbers .... issue ....).

3.3 Final machining takes place on .... using jigs reference .... (**OR** to drawing numbers .... issue.........).

3.4 At this stage goods inspected by .... using .... equipment to confirm correct to specification and is recorded on form shown at Annex ...., to this procedure (**OR** shift log book).

3.5 The items are then sent for weld preparation. As welding (**OR** galvanising, soldering, painting, etc.) is a 'Special Process' and the preparation cannot be checked visually after welding, the preparation is independently checked by .... to .... drawing Issue No. .... with the inspection recorded on .... before the trained welder is permitted to weld the components together.

3.6 After welding, the components are then ground and primed and then painted in accordance with .... Issue ....

3.7 The components are then assembled by operatives trained to assemble .... (**OR** due to the nature of this product special arrangements are needed to handle .... The precautions taken are ....). (**OR** .... plc use a simple but effective method of 'Component Control'. .... are assembled in sets of five. The components to go into the .... are 'trayed-up' onto wooden component control trays with five sets of differently and appropriately sized and shaped indentations in the component control tray. Into the appropriate indentations will be placed five ...., ten ...., twenty ...., fifteen .... The principle being that when 'trayed-up' it is very obvious if a component has been missed. It is also equally obvious at the end of assembly of five .... if a component had been left out.)

Operating Procedure 3: Issue ......

Dated ..../ ..../ 200....

Page 3 of 10

3.8   The final assembled . . . . to be tested on . . . . machine to . . . . pressure for . . . . seconds, with results recorded on form . . . . shown at Annex . . . . to this procedure (**OR** record book/log). Each entry is initialled and dated by the operative. (**OR** at the start of the shift, at least twice during the shift, and towards the end of the shift an additional verification test will be carried out and recorded by the Production Supervisor [**OR** . . . .]).

3.9   A sample of . . . . is taken at approximately hourly intervals by the . . . . These are then independently tested and verified on a different machine to the one above by using the machine in the Inspection Bay. These verification results are also recorded on the form . . . . . and signed and dated (**OR** record book/log). If the verification results do not tally with the production test results the QA Manager (**OR** . . . .) is informed to investigate and resolve the problem. If the results confirm the item is correct the Inspector also checks the visual appearance and ensures any tests required or paperwork specified in the contract is available.

3.10   The items are handled with due care during manufacture with precautions taken for . . . . Also in temporary storage they are carefully stored to prevent damage by working to . . . .

       (*Author's Note: typical examples, anti-static precautions on electronic boards or explosives. Store rubber away from fluorescent lights or electric motors to prevent deterioration due to ozone, etc.*).

3.11   The goods are then packed to drawing number . . . . issue . . . . . . (**OR** . . . . plc Packing Specification . . . . dated . . . .), (**OR** these products [**OR** castings, forgings, ingots, etc.] manufactured by . . . . plc are very robust and hence no packaging or special measures are needed to prevent damage due to handling, delivery or deterioration during storage.) These are paletted (**OR** . . . .) as per the Packing and Shipping notes provided by . . . .
If the contract specifies delivery requirements these are checked and complied with.

3.12   Any requirements for specials or 'one-offs' will be detailed on the 'Temporary Work Instruction Form' (See Annex F). These may be written in ink in manuscript as long as handwriting is legible. The instructions are listed and detailed in a log with the numbers issued in sequential order. A copy of all such Temporary Instructions is kept in a lever arch file for a minimum of 15 years (**OR** . . . .) in the QA Manager's (**OR** . . . .) office.

3.13   If defective items are found, they will be dealt with as per Operating Procedure 4, 'Control of Material' and Operating Procedure 7, 'Corrective and Preventive Actions'.

3A. Implementation (Model for Production/Manufacturing shop with a variety of completely different processes and inspections or tests)

3A.1   . . . . plc's range of products necessitate a range of different tests and inspection.

       These are planned very carefully as a new product is introduced and a 'Standing Work and/or Inspection Instruction' is written, authorised and dated, as required, for each task (see Annex D or E for typical examples). For each product a flowchart or 'process map' is produced, authorised and controlled that identifies the different process and inspection points. (See Annex G). As the number of instructions is less than twenty in total, it has been found convenient to control them by simply making them Annexes to this procedure. (See Annexes . . . .) Therefore as a new Standing Instruction is introduced, or an existing one amended, this procedure is amended to show this latest issue. (**OR** as the number of Standing Work and/or Inspection Instructions is large they are controlled in the same manner exactly as procedures, see Operating Procedure 8, with authorisation, issue and control, with a controlled master list of all the authorised and issued instructions).

3A.2 Each product will always have as its penultimate Standing Work Instruction, one on final inspection and verification check against the ORIGINAL CUSTOMER ORDER, or any confirmatory or test required, and to check that the paperwork specified is available. The final instruction will give details of packing, despatch, delivery advice note or shipping requirements.
If the contract specifies delivery requirements these are complied with.

3A.3 Any requirements for specials or 'one-offs' will be detailed on the 'Temporary Work Instruction Form' (See Annex F). These may be written in ink in manuscript as long as handwriting is legible. The instructions are listed and detailed in a log with the numbers issued in sequential order. A copy of all such Temporary Instructions is kept in a lever arch file for a minimum of 15 years in the QA Manager's (**OR** . . . .) office.

3A.4 If defective items are found they will be dealt with as per Operating Procedure 4, 'Control of Material' and Operating Procedure 7, 'Corrective and Preventive Actions'.

3B. Implementation (model for Metal Machining, Fabrication, or Vehicle Body Shop with jobs of small batch sizes or one-offs)

. . . . plc machine parts and weld fabricate to customers specification. The following controls are used:

3B.1 Each job is checked to ensure it has been authorised by contract review and works order number issued.

3B.2 The Workshop Manager (Foreman/Chargehand or Planner) makes out a double-sided 'Travelling Job Card' (**OR** 'Planning Sheet') for the order, shown at Annex A to this procedure. This job card gives outline details of the job and pertinent drawings and specifications. The Manager (or Foreman, etc.) writes down the sequence in which the machining, fabrication and assembly is to take place. It is not necessary to completely fill out the card before it can be issued to the early stages of a job. If certain tasks need subcontracting these also will be shown in the correct sequence and where to be subcontracted to (this will be done in accordance with Operating Procedure No. 2).

3B.3 The travelling job card will specify turning, milling, welding, etc. with the operative allocated to do that element of the job. If necessary the job card will also specify the machines or equipment that must be used. As all the staff are time served craftsmen (**OR**, trained competent operative) only outline details are necessary together with the drawings/specification being made available. If the order requires a 'coded' welder this part of the job will be subcontracted to an approved subcontractor (**OR** . . . . plc have welders who are registered and training maintained and recorded as coded for . . . .).

3B.4   As each process is completed the operative checks his own work and initials and dates the Travelling Job Card. Any rework, repairs or concessions will also be detailed on the Travelling Job Card by . . . . and signed and dated.

3B.5   At appropriate stages the Manager (planner, etc.) may specify additional verification checks after critical processes. Independent checks will also be specified before an operation on critical or special processes to ensure that the preparation was correct before the process e.g. welding, painting, enclosed assemblies, etc. These will also be initialled and dated on the Travelling Job Card, by the inspector. It is also important that where at all possible checks or measurements are done on different gauges or instruments than those used by the manufacturing operatives.

3B.6   (**OR** if the tolerances are less than ± . . . ., or if the dimensions identified by the planner are critical, the test gauge or instrument used **will** be recorded on the Travelling Job Card.)

3B.7   All final assembly tests and any packaging requirements are specified on the Travelling Job Card. If necessary one works order may have several Travelling Job Cards.

3B.8   Before the job is cleared for despatch the Inspector will give the job one last overall check and will check against THE ORIGINAL CUSTOMER ORDER, and that the 'Travelling Job Card(s)' is/are complete and fully signed-off. The Inspector will then raise any Certificate of Conformity or Special Test Certificate if required and sign and date to complete the Travelling Job Card. If the contract specifies delivery requirements these will be complied with.

3B.9   The Travelling Job Card will be returned to the Works Manager's Office and will be stored for at least 15 years (or longer if a contract requirement). The office will then raise the necessary despatch notes and instructions.

3B.10 If defective items are found they will be dealt with as per Operating Procedure 4, 'Control of Material' and Operating Procedure 7, 'Corrective and Preventive Actions'.

3B.11 Welding and other consumables will be stored as required by the manufacturer's instructions. If they need to be stored in a cold or warm environment it will be provided.

3C. Implementation (model for Site Contractor, Landscaper, Electrical Installation, Construction, Plastering, Painting, Office Cleaning, etc.)

. . . . plc maintains . . . . and installs (**OR** landscapes, **OR** builds, **OR** paints/ decorates/cleans) to customer's specification or contract requirements. The work is usually done on site with materials being supplied via . . . . plc depot. The following controls are used:

3C.1   All of . . . . plc contracts are reviewed as per Operating Procedure 1. After completion of review, these are then issued to the site foreman (**OR** supervisor) with a . . . . plc Job Number. No work can commence on any site without the authority

Operating Procedure 3:   Issue . . . . . .

Dated . . . . / . . . . / 200 . . . .

Page 6 of 10

of such a Job Number. Before issuing the Job Number the Contract Manager will arrange access to the site and, if required, storage and other facilities.

3C2   The vast majority of . . . . plc customers specify exactly what is required by a formal specification with accompanying drawings. These are booked in and recorded in, according to Operating Procedure 8. An additional copy is taken and issued to the site foreman as an information job envelope file as per Operating Procedure 8. Any amendments to the contract are reviewed and booked in and issued in exactly the same way. All the copies issued to the site foreman will be stamped with an official . . . . plc stamp and signed and dated to authorise issue.

3C.3   In order not to have a separate tier of document control to the site foreman with transmission sheets, etc., all amendments to the site foreman's information pack will be personally changed by the Contracts Manager, who will explain the changes required to the Site Foreman.

3C.4   In cases where the client does not issue a specification, ... plc will complete a double-sided Work Specification Sheet, see Annex B, with a plan/sketch of the area if necessary see Annex C.

3C.5   It is common in . . . . plc type of work for the client to request minor amendments to work or additional work during visits to the site, (**OR** emergency call outs). This will be detailed by the Site Foreman in the 'Minor Works Variation Duplicate Book' or site diary and the Site Foreman will sign and date and the customer will be requested to sign, before the minor work (or alteration) is undertaken. The Minor Works Variation Duplicate Book has the pages pre-marked in serial number order. The perforated copy page of the Minor Works Order/Variation will be given to the Contracts Manager to check and arrange additional checks and invoicing, as appropriate. (**OR** A feature of . . . . plc work is occasionally we cannot get into some areas to clean [**OR** . . . .] as per the agreed schedule. . . . . plc have introduced on each site a 'Customer Communication Book' with sheets headed as per Annex . . . . The main purpose is that the customer can leave notes, additional requirements or complaints for our site supervisor to note. If we cannot clean [**OR** . . . .] an area this will be noted, signed and dated, on a sheet in the 'Customer Communication Book' by the supervisor, with the reasons why the area cannot be cleaned [**OR** . . . .] e.g. room locked, occupied, other contractors working, etc.). (**OR** Site work is issued on Job Tickets detailing the work and materials required. At the completion of the job the . . . . will request the customer to sign the Job Ticket to accept the work. This is then sent/given to the Contract Manager to record, verify and arrange to invoice.)

3C.6   The Contracts Manager will maintain for every contract a separate Contract File identified by the . . . . plc Job Number, Client and Site. A copy of every document, letter, specification, minor variation/order must be placed within this contract file. (**OR** in addition to the Contract File there is maintained a Contract Invoice file for every contract to ensure jobs are correctly invoiced and paid.)

3C.7   As the Contracts Manager visits each site he will maintain a hardback A5 diary or log with details, dates of visits, progress on the job, critical inspection of . . . . plc and our subcontractors' work, persons met, clients representative, discussions held, emergency work, injuries, items stolen, conflict of work priorities, effects of adverse weather, or damage by other site contractors, etc. Each day's entry will be signed or

Operating Procedure 3:   Issue . . . . . .

Dated . . . . / . . . . / 200 . . . .

Page 7 of 10

initialled by the Contracts Manager (**OR** Log maintained by Resident Site Foreman, etc., etc.). (**OR** the Contracts Manager [**OR** Site Foreman] will produce a formal, signed and dated report at the end of each week, detailing . . . .]. This will always include a separate sheet recording all site inspections.) This is a vital record both for the Quality System and to provide business records and history of client queries or complaints. It therefore should be maintained neatly and legibly.

3C.8 In appropriate special cases the Contracts Manager may request the client's representative to sign the Diary (**OR** Log) to record visits, key events or inspections. Any inspection prior to a Special Process (e.g. before plastering, inspection of foundations prior to pouring of concrete, inspection of formation levels before top soil added, tree pits before planting, checking critical undercoats or primers before applying first gloss coat, etc.) will be recorded and signed and dated in the Contracts Manager's Diary/Log. Some jobs are so critical or difficult to document and specify that the work will be carried out under the personal supervision of the Contract Manager (**OR** . . . .). Such personally supervised jobs are recorded in the Contract Manager's Diary/Log.

3C.9 Formal inspection prior to . . . . plc taking over a site, pre-contract meetings, pre-hand overs or formal handing over a site back to the client will be recorded on a Formal Visit report (see Annex C, Operating Procedure 1). Where appropriate, a copy will be sent to the client by Recorded Delivery. Alternatively, if the client provides a satisfactory visit report, minutes of meeting or Certificate of Completion, these will be initialled and dated and filed.

3C.10 The day-to-day site work is carried out by tradesmen (**OR** operatives, **OR** . . . .) working under the personal directions or supervision of the site foreman, using the information supplied.

3C.11 The process of . . . . will only be carried out by trained . . . . (**OR** weed killers/pesticides, trained operatives, **OR** testing of electrical installations/equipment, by designated competent persons, etc.). All such applications (**OR** tests) to be recorded in the designated log see Annex . . . .

3C.12 If during an installation or site work, asbestos or material that may be asbestos is discovered, work must stop immediately and report to the manager.

(Author's Note: could be any safety critical features. Could be encountering protected species such as bats, badgers, etc., not a Quality ISO 9000 requirement, but a sensible item to include to show due environmental care and diligence).

3C.13 The spares and equipment are stored in the . . . . plc storehouse. All the items are identified and controlled by the . . . . plc storeman. These are detailed in the Standing Work Instruction . . . . (**OR** Procedure No. . . . .). Also included in this stock is the stock maintained in the Service Engineers' Vans. A list of stock that must be held, and that which is optional is issued by . . . . As part of the Internal Audit schedules every Service Engineer's van is checked with its associated stock and equipment. The precautions when moving . . . . to site are . . . . .to prevent damage (**OR** the installation of . . . . needs special arrangements to prevent damage through static discharge [**OR** . . . .]. The precautions taken are . . . .)

Operating Procedure 3:   Issue . . . . . .

Dated . . . . / . . . . / 200 . . . .

Page 8 of 10

3C.14 If defective items are found they will be dealt with as per Operating Procedure 4, 'Control of Material' and Operating Procedure 7, 'Corrective and Preventive Actions'. (**OR** approximately 3 months after the handover a site/installation the customer is sent a Satisfaction Survey Form [see Annex . . . .] to see if they are still satisfied with the job and to check on the service and work provided, attitude of staff, etc.).

3C.15 (**OR**. It is vital to prevent cross-contamination on cleaning contracts. Toilets will be cleaned using red cloths, red rubber gloves and red buckets, washroom sinks will be cleaned using orange cloths and, orange gloves and orange buckets, kitchen sinks will be cleaned using green cloths, green gloves and green buckets/bowls. These will be segregated during over-night storage.

Food preparation areas will be cleaned with disposal pre-treated disinfecting cloths specifically provided.

(Authors note: above based on traffic light codes. Other codes may be green-clean, pink-sink, blue-loo.)

3D. Implementation (Model for a well regulated industry, profession or service, including approved agents for Ford, Konica, Apricot, etc., etc.)

3D.1 . . . . plc provides . . . . to . . . . throughout . . . . We will abide exactly with the documented procedures laid down in specification (**OR** guidelines **OR** . . . .) . . . . issued on the . . . . date, by the . . . ., with the following exceptions. The industry guidelines call for two copies of . . . . to be supplied to the customer. . . . . plc will always issue three. (**OR** The industry guidelines stipulate test, only after full installation of . . . . It is the policy of . . . . plc that after even the smallest modification to . . . ., the system will be re-tested by an experienced and competent . . . . and a . . . . plc test certificate issued.)

(**OR** The Code of Cremation Guidelines of . . . . stipulate that metal residues will be disposed of according to the directions of the Cremation Authority. It is the policy and practice of . . . . Crematorium that such residues will be buried in a cemetery plot discreetly set aside only for this specific purpose.)

(**OR** . . . . plc will supply, install and maintain photocopiers exactly in accordance with the latest manuals of . . . . Corporation. The . . . . Corporation is the designer and manufacturer of the range of photocopies (**OR** PC's, **OR** X-Ray machines, **OR** . . . .) that . . . . plc provide, the manuals are very comprehensive and cover all the details necessary for the trained technicians of . . . . plc to install and maintain each machine. It is necessary to run the machine to set up and prove or verify the performance of the machines. It is also common practice in this industry to provide machines on trial or return basis or hold some machines on display and trial in the showrooms. If machines are returned the clock is reset to zero, as the invoicing is based on usage. This is often uncontrolled in our competitors and may happen several times with a so-called 'new' machine having clocked many hours of use. It is the policy of . . . . plc that if a machine goes beyond one sixth of the manufacturer's recommended annual usage for that particular model it will be designated and sold as 'pre-used'.)

(**OR** . . . . plc is a Ford main dealer. All our staff are trained and qualified Motor Technicians holding, as a minimum qualification . . . . Apprentices or Trainees will only work on customer's cars under the direct supervision of a designated Motor Technician who will counter-sign their checklists. . . . . plc

Operating Procedure 3: Issue . . . . . .

Dated . . . . / . . . . / 200 . . . .

Page 9 of 10

work using the latest issue of Ford Manuals and checklists for each model, which are maintained and controlled documents as per Operating Procedure 8. In addition to the above the following applies at . . . . plc:

(a) The technical receptionist will walk around every car brought-in for service and repair and will note, with the customer's agreement, any existing scratches, dents or damage to the vehicle on the acceptance paperwork.

(b) Any loose change, jewellery, magazines, papers, etc. found in the car during service or repair will be put into a . . . . plc envelope, sealed and placed on the passenger seat.

(c) If requested, the Motor Technician will meet the customer and explain any features of the service or repair.

(d) Additional controls introduced by . . . . plc for various situations are:

. . . . form/checklist, See Annex . . . . .

. . . . form/checklist, See Annex . . . . .

(e) If the contract specifies delivery requirements, these are complied with.

3D.2 Any requirements additional standing instructions for . . . . plc will be issued and recorded on the form shown at Annex D. Requirements for specials or 'one-offs' will be detailed on the 'Temporary Work Instruction Form' (See Annex F). These may be written in ink in manuscript as long as handwriting is legible. The instructions are listed and detailed in a log with the numbers issued in sequential order. A copy of all such Instructions is kept in a lever arch file for a minimum of 15 years (**OR** . . . .) in the QA Manager's (**OR** . . . .) office.

3D.3 If defective items are found they will be dealt with as per Operating Procedure 4, 'Control of Material' and Operating Procedure 7, 'Corrective and Preventive Actions'.

3E. Implementation (Model for organisations who already have adequate and formally documented process control procedures)

3E.1 The process of . . . . has been documented and the controls implemented for over . . . . years. These were reviewed on . . . . date, to ensure these were current and correct. These control documents are formally known as the . . . . (**OR** listed at Annex . . . .). They have now been formally authorised, re-issued with the page numbering, date of issue and other controls stipulated in Operating Procedure 8. These will now be formally reviewed, audited and maintained within the overall QA system. The process control documents . . . . also include adequate controls and instructions to define the measuring and inspection required and also controls for handling and storage.
If the contract specifies delivery requirements, these are complied with.

3E.2 Any requirements for specials or 'one-offs' will be detailed on the 'Temporary Work Instruction Form' (See Annex F). These may be written in ink in manuscript as long as handwriting is legible. The instructions are listed and detailed in a log with the numbers issued in sequential order. A copy of all such Temporary Instructions is kept in a lever arch file for a minimum of 15 years (**OR** . . . .) in the QA Manager's (**OR** . . . .) office.

3E.3 If defective items are found they will be dealt with as per Operating Procedure 4, 'Control of Material' and Operating Procedure 7, 'Corrective and Preventive Actions'.

Operating Procedure 3: Issue . . . . . .
Dated . . . . / . . . . / 200 . . . .
Page 10 of 10

.... PLC TRAVELLING JOB CARD (OR PLANNING SHEET)

Job No: ....   Client: ....

Description: ....

Requested delivery date: ....

Amendments issued: Original .... / ....   A ..../.........

　　　　　　　　　　　　　　　　Sign　　　　Date

　　　　　　　　　　B .... / ....   C ..../.........

Specification	....	Issue	Date ....
or	....	Issue	Date ....
Drawing	....	Issue	Date ....
Numbers	....	Issue	Date ....

Material cutting details, Material Spec	Diameter/length	Number:	Sign:	Date:

A draft model or workbook to prepare all your required Operating Procedures

.... PLC TRAVELLING JOB CARD (PLANNING SHEET)

JOB DETAILS

Step No.	Process/inspection	Operative	Sign	Date	Identity of test gauge/ instrument

Annex A Oper. Proc. 3:   Issue . . . . . .

Dated . . . . / . . . . / 200 . . . .

Page 2 of 2

.... PLC WORK SPECIFICATION

Job No: . . . .   Client:. . . . .

Client site contact: . . . .   Tel/Fax No. . . . .

Client office contact: . . . .   Tel/Fax No. . . . .

Site: . . . .

Specification issued by . . . .   date . . . .

(Signed)

Page . . . . . . of . . . . . .

Description of work element	Materials required, equipment to take/ use, specification of work, standard to be achieved/inspection required

Annex B Oper. Proc. 3:   Issue . . . . . .

Dated . . . . / . . . . / 200 . . . .

Page 1 of 2

Job No: . . . .   **Client:** . . . .

Signed: . . . .   **Dated:** . . . .   Page . . . . of . . . .

Annex B Oper. Proc. 3:   Issue . . . . . .

Dated . . . . / . . . . / 200 . . . .

Page 2 of 2

.... PLC PLAN/SKETCH

Job .... Site ....

Signed .... Dated ....

Annex C Oper. Proc. 3:   Issue ......

Dated .... / .... / 200 ....

Page 1 of 1

STANDING WORK INSTRUCTION NO. . . . .

For the manufacture and production test of widgets

Specification:

Widgets Mark 4 to . . . . plc   Drawing No.   . . . .   Issue   . . . .

Material:

(i)   As specified on above drawing. To be drawn from . . . . store using withdrawal note ref: . . . .

(ii)   Grease type . . . . from consumable stock.

(iii)   Packing material fibreboard boxes with silicon gel bags, identified as . . . .
Note: the pack is customer supplied material and is kept in . . . . Withdraw using note ref . . . .

(iv)   Tape, Adhesive Type . . . .

Equipment:

Saw No.   . . . .
Lathe No.   . . . .   or No. . . . ., or No. . . . .
Miller No.   . . . .
Hand file.
Go gauge   . . . .
Go gauge   . . . .

Operatives:

Any fully trained, or semi-skilled . . . . plc operative (**OR** only skilled Fitter and Turner, **OR** . . . .)

Process:

1. Have correct copy of drawing to hand, from Foreman's office.
2. Check equipment is available and in working order.
3. Draw enough material to satisfy works order or to complete a day's work whichever is the smaller.
4. Saw to specified length plus approx. $\frac{1}{4}$" to allow for facing either end.

```
Annex D Oper. Proc. 3:   Issue . . . . . .

          Dated . . . . / . . . . / 200 . . . .

                           Page 1 of 2
```

5. Turn in one of the specified lathes to drawing dimensions.
6. Mill in specified machine to drawing.
7. Remove rough edge.
8. Grease the sliding face as indicated on drawing.
9. Visually inspect, test the items on the two 'go' gauges provided and pack into free issue pack supplied but do not seal.
10. Handle with care and attention. Any damaged or suspect items to be placed in the hold tray provided.
11. After the inspector has done a verification inspection and test on sample as per the separate Standing Inspection Instruction, seal all the containers with tape provided. The tape to be length ways equidistant across the gap with $1\frac{1}{2}$" minimum fold down on the sides. Seal top and bottom. Put to one side into piles of 50.

Authorised . . . . Dated . . . .

Annex D Oper. Proc. 3:   Issue . . . . . .
Dated . . . . / . . . . / 200 . . . .
Page 2 of 2

STANDING WORK INSTRUCTION NO. ....

For the testing of the Elite Actuators

Specification:

None required.

Material:

Elite Actuators held awaiting test and inspection.

Equipment:

Actuator test equipment in middle shop only.

Operatives:

No restrictions.

Process:

1. Pick up Elite Actuator.
2. Remove covers from holes.
3. Pull actuator rod up and down to ensure free moving.
4. Place in test ring as per manufacturer's instructions (photocopied in plastic envelopes on wall by machine).
5. Put on test for minimum of 5 minutes (automatically set by machine). (Reference test specification ....)
6. Take off actuator, wipe clean with cloth, replace covers into holes.
7. Check print-off from test machine ensure all readings are above ..... Note actuator serial number onto print-off, place in envelope for today's readings.
8. If reading below ...., set to one side in hold bay. Write serial number on the print-off and place with unit.

Authorised .... Dated ....

```
Annex E Oper. Proc. 3:   Issue ......
         Dated ..../..../ 200....
                          Page 1 of 1
```

TEMPORARY WORK INSTRUCTION NO: . . . .

Product: . . . .   Model/Type . . . .

Contract   . . . .   Issue . . . .   Date . . . .

Drawings   . . . .   Issue . . . .   Date . . . .

Specifications . . . .   Issue . . . .   Date . . . .

Location of Process or Inspection: . . . .

Authorised to be carried out by: . . . .

Special tools/gauges: . . . .

Instructions:

. . . .

. . . .

. . . .

. . . .

. . . .

. . . .

QA/inspection requirements:

. . . .

. . . .

. . . .

Authorised . . . .   Date . . . .
                    Production Manager/Supervisor

Approval QA Manager: . . . .   Date . . . .

Limitations: Only for Contract/Batch . . . .

Until the date of . . . .

A draft model or workbook to prepare all your required Operating Procedures

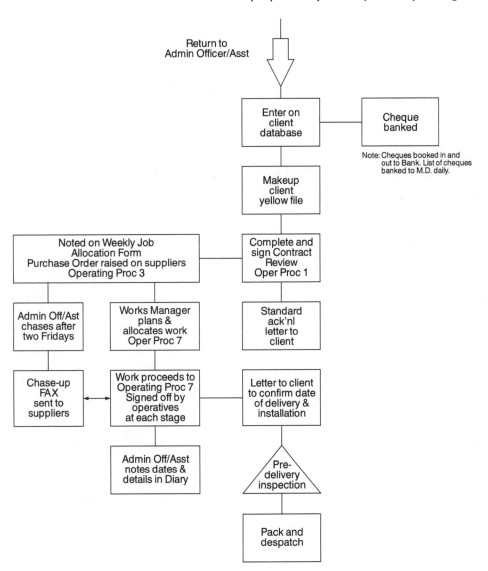

Return to
Admin Officer/Asst

Enter on
client
database

Cheque
banked

Note: Cheques booked in and
out to Bank. List of cheques
banked to M.D. daily.

Makeup
client
yellow file

Noted on Weekly Job
Allocation Form
Purchase Order raised on suppliers
Operating Proc 3

Complete and
sign Contract
Review
Oper Proc 1

Admin Off/Ast
chases after
two Fridays

Works Manager
plans &
allocates work
Oper Proc 7

Standard
ack'nl
letter to
client

Chase-up
FAX
sent to
suppliers

Work proceeds to
Operating Proc 7
Signed off by
operatives
at each stage

Letter to client
to confirm date
of delivery &
installation

Admin Off/Asst
notes dates &
details in Diary

Pre-
delivery
inspection

Pack and
despatch

Process Flowchart for . . . .

J. Bloggs Engrs

FPD Oct. 95 Page 3 of 7

Authorised . . . .

Date . . . .

Annex G Oper. Proc. 3:   Issue . . . . . .

Dated . . . . / . . . . / 200 . . . .

Page 1 of 1

. . . . plc
OPERATING PROCEDURE NO. 4

The procedure for control of materials including non-conforming product

Authorisation and Amendment Record

Issue No.	Date of issue	Prepared by	Authorised by:
1	6 Feb 20	A.N. Other	Initial Draft
2	12 Feb 20	A.N. Other	2nd Draft
3	20 Feb 20	A.N. Other	3rd Draft
4	10 Mar 20	A.N. Other	Informal Issue
5	5 Aug 20	A.N. Other	. . . .
			Signed: M. Director
6			
7			

Operating Procedure 4: Issue . . . . . .

Dated . . . . / . . . . / 2000 . . . .

Page 1 of 7

.... plc
OPERATING PROCEDURE NO. 4

The procedure for control of materials including non-conforming product

1. Purpose and scope

(a) Any items, component, subassembly or raw material that is to be incorporated into .... plc product can be readily identified to type, drawing or part number. That items of similar appearance cannot inadvertently get mixed or substituted.

(b) There is no requirement or advantage in providing any form of traceability for .... plc products/services. If there was a particular contract requirement to do so, special controls would be introduced.

(**OR** where it is a contract or legal requirement, items or material will have traceability to the raw material certificate, **OR** foundry certificate, **OR** forge certificate, **OR** material analysis, **OR** test certificate, **OR** certificate of conformity, etc.) that was provided by the original supplier, (**OR** to the supplier/date of delivery/order No.)

(**OR** where it is a contract or legal requirement, ... plc products [**OR** the following models ...., ..., ..., ...] are manufactured and tested into identifiable lots or batches. These will be supplied with appropriate documentation pertaining to the identity and test results of the individual lot or batch.)

(**OR** .... plc have identified that the ...., ..., ... and .... ranges of models [**OR** food products, **OR** electric fans, **OR** ....] are safety critical [**OR** are susceptible to failure on one component i.e. ....] and will be assembled, [**OR** processed **OR** manufactured] and supplied as identifiable batches and lots [**OR** with lot traceability]. This is in order that if a potential problem batch is identified by customer complaint or return, ... plc can arrange a withdrawal and return of the suspect batch.)

(c) .... plc will ensure an adequate system for inspection Status, i.e. can readily identify whether materials, items, sub-assemblies or final products are good, bad, suspect (**OR** on hold, **OR** quarantined) or reject (**OR** scrap).

(d) When items are identified as suspect (**OR** placed on hold/quarantine) that there is a prescribed and controlled procedure to ensure they are reviewed and controlled through rework, repair or disposal.

(e) Where items (**OR** materials **OR** services) are supplied by the customer to assist .... plc manufacture the final product (**OR** supply the required service) that these are positively identified to prevent misuse, substitution, loss or deterioration.

Operating Procedure 4:   Issue ......

Dated ..../..../ 2000 ....

Page 2 of 7

2. Responsibility

It is the responsibility of the Director and all managers to ensure this procedure is implemented fully and effectively. The supply of a defective product from . . . . plc can have a dramatic effect on customer confidence (**OR** cause fatal accident, **OR** . . . .), this procedure must therefore be enforced rigorously. All staff and operatives with . . . . plc must follow this procedure. Any member of . . . . plc staff or operatives can and must place items under temporary 'hold' if they are concerned that they do not conform to the required specification.

Items under temporary 'hold' can be released only by the operative who placed them on hold, or any manager. Items under formal 'quarantine' or 'rejected' can only be released by the Quality Manager or the Director (**OR** . . . .) and then only by completely the prescribed records. No member of the Production Management have authority to release items from 'quarantine' or 'reject'. Staff and operatives will ensure there are sufficient labels attached to clearly identify the product, or where appropriate its inspection status. As a pallet or container becomes empty all staff shall ensure that redundant labels are removed. The Quality Manager is responsible for re-ordering the Inspection Status labels. If they are not available for a short period temporary labels may be used from white cards.

3. Implementation (typical Production Manufacturing Shop or Site Installation [OR Maintenance])

3.1   Control of Identity
The majority of . . . . plc products are quite unique in size, shape, etc. and cannot be mixed or substituted.

The only items that can get confused are:

. . . .   and   . . . .
. . . .   and   . . . .
. . . .   and   . . . .

Only these items will have an identity number stencilled on the part (**OR** label tied, attached or adhered, etc.).
(**OR** whilst . . . . plc range of products are different in size and shape it is possible for an inexperienced operative to get them mixed. In order to prevent this happening and as an aid to training a series of boards are conveniently placed at the middle of the production shop with an example of each part affixed to the board and identified. By placing an item by the board, the part number of any part can be very quickly identified.)
(**OR** once . . . . plc are placed into sub-assemblies or onto the specified production lines and become work within a specific process they cannot be mixed or substituted. Hence only the raw materials and components are identified. This may be by individual label or part number on the larger products, but the majority are identified to type by labels on the tote bins, pallets or bags).
(**OR** the items made by . . . . plc are one-offs or small batches. With the job is a 'Travelling Job Card' (**OR** 'Planning Sheet, **OR** . . . .)detailing the work to be carried out. This card also identifies the drawings with the items and appropriate material identity. If the identity of any material is not clear it will have a small identity

| Operating Procedure 4: | Issue . . . . . . |
| Dated . . . . / . . . . / 2000 . . . . |
| Page 3 of 7 |

card/label attached or marked directly onto the item, with permanent ink/paint). (**OR** some of the products made by . . . . plc are safety critical. Therefore a formal controlled list of safety critical items is maintained. All of these items will have the identity number marked onto the item, a batch number and also motif to identify the supplier of that component and the year/month of manufacture. This is specified as a requirement of our Purchase Order onto our suppliers. This level of control is essential both to enable product withdrawals and also to throw the product liability [consumer protection] back onto the original supplier.)

3.2   Control of Material, including non-conforming product
Within . . . . plc we have introduced a simple coloured card/labelling system to identify the inspection status of components.

(a)   A yellow label identified as 'hold' see Annex . . . . . . This is a temporary label for use by any member of . . . . plc staff, or for an operative to be able to use, where items are considered suspect or awaiting test or awaiting further components, coat of paint, check by customers representative, awaiting goods inward clearance, awaiting suppliers certificate of conformity, confirmation of customer's specification, formal order documentation, customers returns awaiting checking, etc. Items can be placed on 'hold' by any member of . . . . plc staff. It can be removed only by the person placing it into 'hold', or any . . . . plc Manager or Director.

(b)   An orange label identified as 'quarantine' see Annex . . . . This is where items are confirmed as not to the required specification. They must then go through a simple but formal review procedure before they can be reworked, repaired, scrapped, etc. (See Operating Procedure 7). Any member of . . . . plc staff can quarantine items. It can only be lifted by the QA Manager or the Director (**OR** . . . .) with the appropriate completed authority and record forms.

(c)   A red label identified as 'reject' see Annex . . . . . . These items are not suitable for . . . . plc products; the label will not be removed whilst the items are within . . . . plc. They will be sent for disposal, or returned to the supplier with the red 'reject' label securely fixed. If appropriate instructions may be given to 'ruin' or 'slight' the product by painting or spraying red (**OR** chiselling across the working face **or** stamping 'R' onto working face **OR** chopping off the connecting wires, **OR** . . . .) to positively prevent their re-introduction into the manufacturing process. A 'Reject' label can only be fixed by an approved Inspector, a Manager or a Director.

(d)   A blue label identified as 'special' see Annex . . . . . . It will generally only be authorised by the Goods Inward Inspector, the Quality Manager or the Director. This label will be used for the following:

* to positively identify 'free issue' or customer supplied products to prevent its misuse, substitution or loss. At the completion of the order/contract all excess 'free-issue' material must be returned to the customer. It will only be disposed of following written instructions or permission from the customer.
* to identify all special batches where lot traceability has been requested.
* to identify items or batches that are special, have unique properties and must not be mixed with normal production.

Operating Procedure 4:   Issue . . . . . .

Dated . . . . / . . . . / 2000 . . . .

Page 4 of 7

* special marking or assembly trials or items under development; also can be used to identify 'first offs' for critical examination or machining trials.
* for use to identify test supplies taken by laboratory staff.
* for use to identify 'Approved Visual Standards'. The inspection of item .... is a subjective visual inspection. To provide guidance to acceptable standards on the features, acceptable surface defects, cosmetic appearance, etc. acceptable examples of product are provided. These are clearly identified and authorised using the label above, (**OR** there are three examples of this approved visual standard. All three have been approved, signed and dated by a representative of the customer and the Quality Manager [**OR**.......] of .... plc. The customer has selected and holds one, the Quality Manager holds another and there is one available at the place of work). Care should be taken not to damage the visual standard. If it deteriorates it should be replaced. The blue label will stay on the item whilst it is within .... plc.

(e) (**OR** .... plc machines from various diameters of metal and [**OR** plastic) bar stock that are held in racks in the bar stores. It is vital that these are readily identifiable to type and specification of material. Hence wherever practical the ends of the bar will have a colour code painted onto the end. Care must be taken when doing this as some stockists have their own colour codes. The colour codes are shown at Annex .... at the end of this procedure. A copy of the latest issue of this annex will be displayed in a plastic cover on the wall in the bar stores and in the machine shop. This will be personally controlled and updated by the Quality Manager to ensure it is the latest issue. Care must be taken by machinists when sawing bars that the material identifying code stays with and on the bar left in stock. If there is a usable 'cut-off' that can be returned to store the code should be repainted onto the end. All material without the paint code, such as angles, plate must have the material identified on the item by permanent labels or by 'paint stick'. Material ordered or set aside for specific jobs will have the job number affixed or painted onto it.)

(**OR** the majority of bar used by .... plc is general purpose mild steel spec. .... This will not be paint coded, all other material will be carefully coded.)

(**OR** useable bar stock off-cuts are put in the bin [**OR** shelves, etc.] marked 'Unidentified and uncontrolled material'. This material is to be used for machine setting up or development trials. It may be used on low cost, low duty jobs [where the customer has not specified specific grade or material] on the instruction of the Director or Quality Manager [**OR** Project Engineer, **OR** ....] this will be detailed on the Job Card records and initialled and dated by the Director [**OR** ...., **OR** ....].)

(f) A green label identified as 'Accepted/Passed Test' see Annex .... This is an OPTIONAL label. Due to effective controls introduced within .... plc, it can generally be assumed that any items without a hold, quarantine or other specified label is in fact either acceptable material or work in progress. It may be used to very clearly identify batches or items that have been subject to critical examination or test and have passed. It may also be specified on individual Job Cards to be attached to each item or batch for certain Quality Critical products (**OR** batches to be supplied to Sony, Hitachi and Aiwa will always be subject to a formal verification inspection and a .... plc green label neatly completed and fixed to each item or packing box).

Operating Procedure 4:   Issue ......

Dated ..../..../ 2000 ....

Page 5 of 7

(**OR** . . . . plc products are supplied as home aids to the handicapped. The customer particularly needs assurance that the product has been carefully tested hence every product will have a . . . . plc green label attached. The label will be NEATLY completed and signed off with an identifiable signature). The green label will only be affixed and signed by the operative or member of staff either inspecting or testing the product.

(g) (**OR** . . . . plc wishes to positively identify items that have been the subject of a concession/waiver or have been sold as sub-standard/seconds. Such items will be marked (**OR** stamped) with a 'C' or 'S' as appropriate.)

(h) A white label identified as 'Uncontrolled'. This is an optional general-purpose label that may be attached if appropriate to items that are outside the QA control system. Can be attached by any member of . . . . plc staff. Typical examples are competitors' products, drawing office samples or exhibition pieces. This label will generally stay on the item whilst it stays within . . . . plc.

3.3 Wherever possible or practical items that are under 'quarantine' or 'rejected' will be segregated to a designated area. In some cases due to size of product (**OR** explosive building limits **OR** chemical hazards, **OR** . . . .) it is not possible to physically segregate. In these cases special care should be taken to ensure each item or container is clearly identified as 'quarantine' or 'reject'.

3.4 In some cases it is essential (either to meet client's critical delivery date, or to keep operatives productively employed) to work on material under 'hold' or 'quarantine'. If the Works Manager or Director wishes to take this risk of subsequent rejection, they must authorise its use by a signed and dated memo. The items must retain their identity and inspection status as 'hold' or 'quarantine' until the items are subsequently accepted or rejected.

3.5 All entries onto labels, both identity and inspection status, must be clearly legible and in **permanent** ink, paint, engraving, moulded in, stamped, etc. (**OR** as an additional improvement to labelling items, where specified it is permitted to use a . . . . plc approved workmark to confirm items have been personally inspected and are accepted. This is a rubber stamp [**OR** a metal punch]. These are shown on Annex . . . . . The Quality Manager has a record book identifying the unique workmark number to the member of staff, date of issue and signature and date of withdrawal. To prevent any conflict any workmark withdrawn, is scrapped and not re-issued, also the identity number or initials will never be re-introduced.)

3.6 It has been found that certain members of . . . . plc staff have unusual or illegible signatures. The QA Manager retains a control book of inspectors, operatives and staff who can complete Quality Records. This is known as the squiggle list. It identifies the name of the authorised signatory, with an example of signature and an example of initials.

3.7 Certain items received (**OR** used or made) at . . . . plc have a limited shelf life. As these are received (**OR** made) they will be identified on their . . . . with indelible ink or paint with their withdrawal date (or have a label attached with date of retest, withdrawal, etc.

Operating Procedure 4: Issue . . . . . .

Dated . . . . / . . . . / 2000 . . . .

Page 6 of 7

3.8   When items are identified as 'Quarantined' or 'Reject' they will then become subject to the review and control of Operating Procedure 7.

3A. Implementation (Typical Service Business)

3A.1   . . . . plc does not manufacture products but provides a service. The identity of the specific service being provided is very clear from the documentation (**OR** work order **OR** job requirement **OR** job file, **OR** . . . .) held by the provider.

3A.2   The inspection status of work (**OR** each document) is clear from the progress of signatures on the control documents. From simple examination of the documents it can clearly be identified which item/service has just been issued, which is in progress, which has been completed and signed off and what has been verified. Note must also be taken of Operating Procedure 8, reference 'Control of Contract Information, Drawings and Specifications', with regard to control of mail and other documents. (**OR** to have full control of incoming documents, and also documents between departments or individuals, each document will have small transmittal slip stapled or securely attached [**OR** be stamped with rubber stamp]. This will show Project/Job number, who to see it for information, who to action it, who to file it, with space available for initials and date to confirm it has been seen or actioned.)

3A.3   As an additional Inspection Status control we have a 'Quarantine' label. These can be applied to doubtful or non-conforming files or envelopes, etc. to prevent despatch.

3A.4   Occasionally the customer provides . . . . plc with copies of their specifications (**OR** examples of products, **OR** confidential documents/records). To prevent loss, misuse (**OR** unauthorised copying), these will be identified by a 'Blue' Special Control label see Annex . . . .

3A.5   Within the documents held are old or obsolete documents, brochures, standards, wall charts, etc. These are held for general and background information only. These are identified with a stamp stating 'uncontrolled copy for info only'.

HOLD   Sign . . . .

                                                  Date . . . .

Items: . . . .

Reason: . . . .

                . . . .

QUARANTINED   Sign . . . .

                                                  Date . . . .

Items: . . . .

Reason: . . . .

                . . . .

Agreed actions: (Regrade/Rework/Concession/Repair-Concession/Sift/Scrap/Other)

Sign . . . . Date . . . .

REJECTED   Sign . . . .

                                                  Date . . . .

Items: . . . .

Reason: . . . .

                . . . .

Disposition action:

Sign . . . . Date . . . .

Annex A Oper. Proc. 4:   Issue . . . . . .

Dated . . . . / . . . . / 2000 . . . .

Page 1 of 1

. . . . plc   . . . . plc
                                                                        Motif
                                                                        _____

SPECIAL CONTROL

Items: . . . .

Reasons: . . . .

            . . . .

            Sign . . . .   Date . . . .

. . . . plc   . . . . plc
                                                                        Motif
                                                                        _____

ACCEPTED/PASSED TEST

Items: . . . .

Batch: . . . .
(**or** Serial No.)

            Sign . . . .   Date . . . .

UNCONTROLLED

These items are not part of . . . . plc

products and are outside the QA control system

Annex B Oper. Proc. 4:   Issue . . . . . .

Dated . . . . / . . . . / 2000 . . . .

Page 1 of 1

. . . . PLC WORKMARKS

A.5   Identifies . . . . plc

Unique inspectors number

Annex C Oper. Proc. 4:   Issue . . . . . .

Dated . . . . / . . . . / 2000 . . . .

Page 1 of 1

.... plc
OPERATING PROCEDURE NO. 5

The procedure for control of measuring and processing equipment

Authorisation and Amendment Record

Issue No.	Date of issue	Prepared by	Authorised by:
1	10 Sept 20	A.N. Other	Initial Draft
2	20 Jan 20	A.N. Other	. . . . Signed: M. Director
3			
4			
5			
6			
7			

Operating Procedure 5:   Issue . . . . . .

Dated . . . . / . . . . / 2000 . . . .

Page 1 of 5

.... plc
OPERATING PROCEDURE NO. 5

The procedure for control of measuring and process equipment

1. Purpose and scope

The purpose and scope of this procedure is to ensure that:

(a) all the measuring, checks and tests carried out by .... plc are with instruments that have adequate accuracy and provide reliable readings and results;

(b) process equipment that directly affects the product (**OR** service) quality has suitable maintenance at appropriate intervals;

2. Responsibility

The Directors or Top Management have the responsibility to ensure that all inspection, measuring and test equipment is correctly controlled and 'calibrated'. The Quality Manager has been delegated to control the test and inspection equipment on a day to day basis. The Production Manager (**OR** and Transport Manager as part of our scope of approval is the transport, delivery or distribution of .... plc products/services,) is responsible for identifying production machines that have a critical effect on the quality of .... plc products. The machines (**OR** lorries, delivery vehicles) will be maintained as per formally controlled and authorised maintenance schedules.

3. Implementation – measuring and test equipment

3.1 All measuring and test equipment must be bought by or through the Quality Manager.

3.2 Each individual instrument or test equipment is registered on the Measuring and Test Equipment Register (**OR** Record Cards) shown at Annexes F and B or C (**OR** some skilled staff hold their own simple measuring equipment such as verniers and micrometers. They must also be brought under the control of this procedure, listed in the register, or removed from the site.)

3.3 Each individual instrument or test equipment must have a unique identity. If there is only one model/make of instrument within .... plc no further identity is required. If there are more than one they must be identified by a serial number permanently etched or engraved on the instrument (either the manufacturers or a .... plc serial/ identify number is acceptable). If the boxes/instruments are interchangeable, the identity number must be on the instrument itself and not on the instrument box.

Operating Procedure 5: Issue ......

Dated ..../ ..../ 2000....

Page 2 of 5

3.4   All instruments are initially calibrated at periods shown at Annex A. Individual instruments that may have more use and wear or have more critical application may have shortened times between calibration. Some infrequently used instruments may be labelled and the record card endorsed 'to be calibrated before use'.

3.5   All Measuring and Test Equipment Record Cards are stored in individual, punched A4 plastic bags (**OR** Rexel, or similar, sleeves) and are kept in a lever arch file. When new instruments are purchased an initial calibration record is specified and obtained from the supplier. This is stored behind the card within the individual Rexel, or similar, sleeve.

3.6   During the last two weeks of each month the Quality Manager will review the cards and identify the instruments that are due for calibration in the month after the next. He will obtain quotes as required and place purchase orders for the calibration.

3.7   All instruments and equipment used to check the quality of our product or service, are calibrated at external test houses. By preference these will be either formally recognised approved test houses (e.g. NAMAS, or UKAS, or state approved laboratory/test-house, or ISO 9001:2000 approved organisation with scope of approval for calibration of the particular instruments, or a local Authority Trading Standards laboratory) or may be the original equipment manufacturer or his formally approved agent. If it becomes essential to use an 'unapproved' organisation the Purchase Order will specify that clear evidence of satisfactory calibration is required, with traceability to a national standard if appropriate. All Purchase Orders will specify that a 'Certificate of Calibration' must be supplied. All firms or organisations providing a 'Calibration' service are included on the . . . . plc 'Approved Suppliers List'. (**OR** within . . . . plc we have two levels of instruments 'I' and 'P'. The 'I' or inspection instruments will be calibrated by the outside organisation. By preference the outside organisation will be either formally recognised approved test houses [e.g. NAMAS, or UKAS, or state approved laboratory/ test-house, or ISO 9001:2000 approved organisation with scope of approval for calibration of the particular instruments, or a local Authority Trading Standards laboratory] or be the original equipment manufacturer. If it becomes essential to use an 'unapproved' organisation the Purchase Order will specify the evidence required with traceability to a national standard if appropriate. All Purchase Orders will specify that 'Certificate of Calibration' must be supplied. All firms or organisations providing a Calibration service are included on the . . . . plc 'Approved Suppliers List'. The 'P' or production instruments or gauges are used by the production operatives and Production Department Manager to check the conformity of product to specification. As the products made by . . . . plc are not to critical or tight tolerance it is considered that a full calibration process, with controlled environment, equipment, codes, standards and specially trained staff, is not appropriate for 'P' gauges. The 'I' or inspection instrument will be **only** used by the Quality Manager or a skilled inspector to carry out inspection or verification checks. Also the 'I' instrument will be used to 'check' the 'P' instrument or gauge. The 'P' gauges and instruments will be recorded on a separate 'Production Gauges and Test Equipment Record Card' (see Annex C). The length of time between the checks and the detail of the checks to be carried out, the appropriate tolerances and who is trained and may carry out the checks are specified on the individual cards.)

Operating Procedure 5:   Issue . . . . . .

Dated . . . . / . . . . / 2000 . . . .

Page 3 of 5

3.8 After calibration or checks the results will be reviewed. If it is thought likely from the results that an instrument is likely to go out of calibration it will be withdrawn or the calibration period substantially shortened. (**OR** if appropriate an 'I' gauge can be down graded to a 'P' gauge.) If the results show that the instrument or gauge has gone out of calibration or tolerance whilst it has been used this should be carefully considered by the QA Manager to see if any action is required on possibly out of specification product that may have been made and supplied to a customer in this period. The QA Manager will write and record their comments/action on the Gauge/Instrument Record Card and sign and date. If the action is serious and may result in product being returned or a product withdrawal this must be reported in writing to senior management and included with the Management Review, see Operating Procedure 7.

(**OR** It is known within . . . . plc that certain measurements or tests are critical. On these tests the record card demands the identity of the test instrument for each set of test results. Hence if an instrument goes out of calibration tolerance . . . . plc can quickly identify any batches affected and review the situation and take and record appropriate action.)

3.9 Calibration labels may be affixed to some equipment by outside calibration houses. Such labels are for information only, the control of the equipment is the Record Cards only. (**OR** in order to identify more clearly the calibration status of gauges or instruments, a proprietary label is attached to each item giving next 'calibration due date'.)
(*Author's note: the latter is the more commonly seen. However, do consider: Are you giving the assessor an extra item to cross-check your system against? What if the label falls off or becomes illegible?*)

3.10 None of . . . . plc's testing equipment depends on computer software for correct functioning and hence this element of the standard does not apply (**OR** items . . . . and . . . . depend on computer software for their functioning and accuracy and reliability of measurement. These are included in the calibration system and are calibrated by the manufacturers [**OR** approved agents] to their own specialist procedures).

3.11 The following instruments or equipment are not included within . . . . plc QA control system:

(i) Any gauges supplied as 'free-issue' by the customer to check product to their specification.
(ii) Rules and tapes are not formally controlled, calibrated or checked. These are used for visual measurements and checks, and obviously the user errors in these visual checks far exceed any tolerance error in the rule or tapes. The metal rules and tapes purchased will only be of well-known high grade proprietary brand names and are checked for obvious damage before use. If damaged they will be replaced immediately. Wood rules or cloth tapes will not be used.
(iii) The electrical consumption meters or the pressure gauges on hydraulic presses and on air presses and vessels, are for indication or insurance purposes only and not within the QA system. If considered helpful these may have a label stating 'uncalibrated' attached. (**OR** with the exception of the hydraulic gauges on the . . . . presses, the consolidation and density of the . . . . pellets are in proportion to the hydraulic pressure, hence these gauges are calibrated.) (**OR**

Operating Procedure 5: Issue . . . . . .
Dated . . . . / . . . . / 2000 . . . .
Page 4 of 5

with the exception of the temperature and pressure gauges and the timers in the rubber vulcanised oven, (**OR** . . . .) as these directly affect the product quality they are included within the calibration system.)

(iv) . . . . plc have weighing scales within the despatch bay. These are for calculating postage or lorry charges only and do not affect product quality and are excluded from the QA control system. (**OR** . . . . plc is a secretarial bureau. it is essential that our clients are charged the correct postage, hence the weighing scales are checked by the post office in accordance with the above procedure). (**OR** . . . . plc products depend on correct measurement of weights for mixing correct quantities, hence, all weighing equipment is carefully controlled and calibrated in accordance with the above procedure.)

4. Implementation – process equipment

The Production Manager with the assistance and advice of the Quality Manager have identified critical items of machinery or process equipment that have a direct affect on the continuing process capability of . . . . plc to meet product specification or contract requirements.

These are identified on the list shown at Annex D, with the designated period for suitable maintenance. A record card is shown at Annex E for the purpose of recording that the maintenance has taken place.

In the cases where the card shows that continuous regular maintenance or cleaning has to take place hourly, daily, at the end of each shift, this will not be recorded on this card but will be recorded on a suitable production log or diary.

(**OR** . . . . plc is a fast food outlet where the hygiene must be of a high standard and is clearly visible to the public. The cleaning schedule is shown at Annex . . . . with a record card to show cleaning carried out and by whom is shown at Annex . . . . The record cards for the kitchen equipment will be kept in the crew captain's (**OR** Head Cook's office **OR** . . . .) whilst the record card for the toilets will be displayed in the toilet area.)

(**OR** part of . . . . plc business and ISO 9001 scope of approval, is the transport or delivery of goods. The Traffic Manager will keep a log card (**OR** book) shown at Annex . . . . of when the maintenance is due. It will also be used to record unscheduled breakdowns, accidents or changes of tyres.) The recommended periods between maintenance of items or machines (**OR** vehicles) will be carried out exactly as the original vehicle manufacturer's recommendations.

(**OR** within . . . . plc we have several pieces of portable electrical equipment. Use is made of the same record cards and schedule [**OR** books, etc.] to identify and test these equipments by competent persons in accordance with the Electricity at Work Regulations.)

(*Author's note: unless a contract requirement the above electrical safety check is not a requirement of ISO 9001; however, you might consider it makes sense to include the above and other appropriate legislative items, to ensure they get done. It may also be very useful in cases of accident or claims, to be able to positively demonstrate that you had formally instructed it to be checked, etc.*)

Operating Procedure 5:   Issue . . . . . .

Dated . . . . / . . . . / 2000 . . . .

Page 5 of 5

CALIBRATION OR CHECK PERIOD FOR MEASUREMENT GAUGES, INSTRUMENTS OR EQUIPMENT

The periods below are the maximum length of time between calibration. They may be shortened for individual gauges or instruments in cases of excessive use, or critical application.

An item can be calibrated up to 2 months early but should not go over one month late.

0–6" vernier micrometer (mechanical or electronic models)	6 month
0–larger than 6" verniers	1 year
0–1" gap micrometers	6 month
0–3" gap micrometers	6 month
0–greater than 3" micrometers	1 year
Depth & height micrometers & dial test indicators	6 month
Engineers squares, straight bars, spirit levels, etc.	1 year
Precision slip gauge set (master)	2 year
Precision slip gauge set (working)	6 month
Surface tables	2 year
3 co-ordinate measuring machines	6 month
Theodolites, pedometers, ...	12 month
Mercury thermometers	Calibration on receipt only
Measuring jugs, instruments and tubes	Calibration on receipt only
. . . . equipment. Calibration is not required as is self checking. Service only	12 month
CAD plotter or printer (only where drawings or charts are used directly for marking out or measuring)	6 month
Bourdon tube pressure gauge	6 month
Electrical temperature, pressure gauges or charts	12 month
Lorry tachometers (specified legal requirement is)	. . . . . . month
Voltmeters, ammeters, etc.	6 month
Plug, ring & fixed gap gauges	1 year
Adjustable gap gauges	6 month
Weights & scales	1 year

Annex A Oper. Proc. 5:   Issue . . . . . .

Dated . . . . / . . . . / 2000 . . . .

Page 1 of 1

MEASURING & TEST EQUIPMENT RECORD CARD
(Checked at outside Test House)

Calibration Period = Standard/* . . . . months

Period changed

Reason: . . . .   Sign . . . .   Date . . . .

Months . . . .

Instrument/gauge No: . . . .
(if required)
Description: . . . .

Manufacturer: . . . .

Date of Purchase: . . . .   Original Certificate No: . . . .

Special Calibration requirements: . . . .

. . . .

Date sent	Date calibrated	Test House	Calibration Certificate Number	Results reviewed (Sign)	Date put back into circulation

* delete as appropriate

MEASURING & TEST EQUIPMENT RECORD CARD
(Checked at .... plc)

Calibration Period = Standard/* .... months

Period changed

Reason: .... Sign .... Date ....

Months ....

Instrument/gauge No: ....
(if required)
Description: ....

Manufacturer: ....

Date of Purchase: .... Original Certificate No: ....

Special Calibration requirements: ....

....

The following to be checked by _____

	Detail Checks	Sign/ Date							
(i)									
(ii)									
(iii)									
(iv)									

* delete as appropriate

Annex C Oper. Proc. 5:   Issue ......

Dated .... / .... / 2000 ....

Page 1 of 1

MAINTENANCE SCHEDULE OF . . . . PLC PROCESS EQUIPMENT
(OR AND/OR VEHICLES)

Identity	Description (and location)	Designated suitable maintenance	Designated period, days, months, No. of miles, No. of revs, units of production, etc.	Signed

Annex D Oper. Proc. 5:   Issue . . . . . .

Dated . . . . / . . . . / 2000 . . . .

Page 1 of 1

PROCESS EQUIPMENT (OR VEHICLE) MAINTENANCE RECORD CARD
Equipment description and identity . . . .

Date	Maintenance or other work carried out	Reading on equipment 'clock' if appropriate	Signed

Annex E Oper. Proc. 5: Issue . . . . . .

Dated . . . . / . . . . / 2000 . . . .

Page 1 of 1

MASTER LIST OF ALL . . . . PLC MEASURING AND TEST EQUIPMENT

Gauge No. (if required)	Description	Manufacturer	Date purchased	Date withdrawn

Annex F Oper. Proc. 5:   Issue . . . . . .

Dated . . . . / . . . . / 2000 . . . .

Page 1 of 1

.... plc
OPERATING PROCEDURE NO. 6

The procedure for training

Authorisation and Amendment Record

Issue No.	Date of issue	Prepared by	Authorised by:
1	3 Feb 20	A.N. Other	Initial Draft
2	20 April 20	A.N. Other	T. Boss Managing Director
3	17 Sept 20	A.N. Other	. . . . Signed: M. Director
4			
5			
6			

Operating Procedure 6:   Issue . . . . . .

Dated . . . . / . . . . / 2000 . . . .

Page 1 of 3

.... plc
OPERATING PROCEDURE NO. 6

The procedure for training

1. Purpose and scope
The purpose and scope of this procedure is to ensure that:

(a) all employees' and directors' training needs are reviewed and any training needs identified are provided;

(b) ....plc employees are competent and are always adequately educated, trained or experienced for their job;

(c) training records are maintained for all employees and directors;

(d) (**OR** that .... plc have primary evidence to demonstrate our commitment to training and to the U.K. Government 'Investor in People' (IIP) scheme.

2. Responsibility
The responsibility for implementing the procedure rests solely with the Managing Director (**OR** the managers of each department for their own staff and with the Managing Director in relation to the training needs of the managers and also to formally review the MD's own training needs).

3. Implementation
3.1 As each new employee or director joins .... plc there is a short induction meeting, see Annex A for items to discuss. As part of the ISO 9000 introduction programme all the existing employees went through the same short programme, with Annex A provided for each employee.

The meeting may be carried out by any member of staff who is considered to have enough experience and background knowledge but it is usually carried out by the employee's immediate supervisor or departmental manager. The Induction Programme when completed and signed is retained with the training records.

3.2 For each and every permanent employee or director there is an Employee Training Review and Record Card kept, see Annex B. The Managing Director will complete his own and the Managers' Cards. The remainder will be completed by the Departmental Managers.

(Authors note: Annex B should be seen as a minimum record. Such a form can be made much more useful by adding other valuable information, e.g. Next of Kin, telephone numbers in case of accident. Other qualifications, experience, languages spoken, etc., etc.)

Operating Procedure 6:   Issue ......

Dated ..../..../ 2000 ....

Page 2 of 3

At commencement of employment (or at interview) this card will be completed. If training needs are identified these are submitted for the review, approval and action of:

(a) a Manager for in-house training/experience;
(b) the Managing Director for external training.

On completion of the review it is signed and dated at the top right hand corner. The record and training needs will be reviewed at least annually and will be signed and dated. If a card is completed or full, it will be crossed through and a new one stapled to the top. If a training need or extra experience requirement has been identified it will be identified in the right hand column and if it is essential it will be actioned ASAP. Desirable training/experience needs will be provided as time and budget permits. External training can only be authorised by the Managing Director. Internal training can be authorised by . . . . Training may be by:

(a) specialist training companies or trade associations;
(b) via local college or university;
(c) or on-site by consultant or appropriate member of . . . . plc staff, which may include on-the-job training.
      The training provided will be reviewed at the Management Review Meeting to ensure effective and no further action is required.

3.3 Where formal qualifications, apprenticeships, certificates, diplomas, degrees or licences are claimed or required to demonstrate competence, a photocopy of the original document will be kept with the above card. (**OR** when staff are required to drive vehicles, their actual licences will be called in once per year photocopied and checked for endorsements, etc.)

(**OR** . . . . plc has a large machine shop that has several skilled and semi-skilled operatives. A matrix form has been produced showing the names of the individual operatives, cross referenced to the tasks they are trained and competent to do. Each competence is agreed and signed by both the foreman and the operative.

3.4 Occasionally . . . . plc take on temporary staff as typists or labourers (**OR** . . . .).
      Where these temporary staff duties do not affect the quality of the product/service that . . . . plc supply, no employee Training Review and Record will be created. However, **all staff whether permanent or temporary must complete the induction programme**.

(**OR** . . . . plc has new models introduced at regular intervals. Service Engineers [**OR** . . . .] required to attend product training courses at least one every month [**OR** . . . .]. The Training Manager [**OR** . . . .] will keep a register of attendees.)

Operating Procedure 6: Issue . . . . . .

Dated . . . ./. . . ./ 2000 . . . .

Page 3 of 3

INDUCTION RECORD
TO . . . . PLC

* Introduction to . . . . plc firm & history ....

* Introduction to . . . . plc products/service (stressing the importance and relevance of their activities and contribution) ....

* Introduction to . . . . plc site, departments, canteen, toilets, etc. ....

* Introduction to the specific products of interest to new starters ....

* Check on employee's transport arrangements ....

* Time of start, lunch and finish
Method of clocking in/off (or recording, etc.) ....

* Condition of contract, pay, pensions, insurance, arrangements for sick leave, holidays, etc. ....

* P45 (income tax record form) from previous employer ....

* Management structure of . . . . plc and in relation to employee and immediate supervisor and colleagues ....

* Discipline arrangements ....

* Health & Safety rules and arrangements ....

* Security and confidentiality ....

* Introduction to QA policy and procedures ....

* Introducing new starter to office staff ....

* Introducing new starter to immediate colleagues ....

* Introduction to designated workmate to assist introduction to . . . . plc for first four weeks ....

* . . . .

* . . . .

Trainer . . . . New Starter . . . .
      Signed   Signed   BLOCK LETTERS

Date: . . . .

.... PLC EMPLOYEE TRAINING REVIEW & RECORD

Surname .... First Name .... Start date .... Staff No:	Date	Sign review	Date	Sign review
Current position (i) .... Date .... (ii) .... Date .... (iii) .... Date				

List those required for post	List the employees existing	Identified training needs State whether desirable or essential	Experience/training provided Date/venue/details
(i) Attributes			
(ii) Experience			
(iii) Qualifications			

.... plc
OPERATING PROCEDURE NO. 7

The procedure for management review, internal audits and corrective/preventive actions

Authorisation and Amendment Record

Issue No.	Date of issue	Prepared by	Authorised by:
1	12 Sept 20	A.N. Other	1st Draft
2	14 Oct 20	A.N. Other	2nd Draft
3	2 Nov 20	A.N. Other	3rd Draft
4	12 Dec 20	M. Experience	4th Draft
5	2 Feb 20	M. Experience	. . . .
			Signed: M. Director
6			
7			

Operating Procedure 7:   Issue . . . . . .

Dated . . . . / . . . . / 200 . . . .

Page 1 of 7

.... plc
OPERATING PROCEDURE NO. 7

The procedure for management review, internal audits and corrective/preventive actions

1. Purpose and scope

The purpose and scope of this procedure is to ensure that:

(a) In addition to the normal management reviews of our business, the quality aspects of our business are also effectively reviewed and appropriate actions taken as required.

(b) Every aspect of .... plc activities that affect the quality of the products or services that we provide are audited at least once (**OR** twice) per year. This will ensure that there is no gradual deviation from the required standard and QA system. It also ensures that the various elements of the Quality System is checked for continuing suitability and relevance.

(c) As errors are found, or brought to our attention, in our products, services or management systems, these will be reviewed for immediate corrective actions and also reviewed to see if longer term preventive action is required or it is possible to prevent recurrence.

2. Responsibility

The Managing Director is directly responsible for all these activities (**OR** the planning, organisation, recording and administration have been delegated to the Quality Manager [**OR** .... Quality Consultants have been subcontracted to carry out the internal audits by formal purchase order clearly specifying their duties and responsibilities]. However, the responsibility and actions remain with the Managing Director). All staff have a responsibility to report any customer complaints to their supervisor who must enter it into the prescribed record. All staff have a responsibility to co-operate fully with an approved internal or external auditor.

Operating Procedure 7: Issue ......

Dated ..../ .... / 200 ....

Page 2 of 7

3. Implementation

3.1 Implementation management review

A formal management review is held at least every 3 months (**OR** 6 months **OR** 12 months) with minutes taken and typed, identifying the actions required and progress/completion on any actions noted. These are chaired by the Managing Director, the minutes are signed and dated by the Managing Director and circulated to attendees and those with need to know, or actions upon them.

Prior to the meeting it is essential that appropriate figures are collected, collated and analysed. E.g. scrap percentages, number of defects, number of customersí returns, number of customer complaints or compliments, non-compliances at internal audit etc;

The following agenda will always be followed as it provides a relevant checklist, with notes on each item, including a note to state 'there is nothing to report' where applicable.

(a)# Background information affecting the business e.g. growth, decline, new products etc.;

(b)#~ review of minutes and actions from previous Management Reviews;

(c) Current work-in-progress;

(d) Future order book (confirmed, probable, possible);

(e) Stock levels, raw materials/finished goods;

(f) Marketing and sales activities; competitors and new products/technology/ legislation;

(g) Financial performance against budget/revised target;

(h) Staff and personnel, effectiveness, recruitment, changes, new responsibilities;

(i)# Results of internal Quality Audits, external certification body surveillance visits and second party audits; review need for any area to have additional internal audits;

(j)# Internal scrap, wastage rates, rework, repairs, concessions, improvements in process performance and/or product conformance;

(k) Installation reports;

(l)# Customer complaints, or favourable comments, or feedback; customer survey reports;

(m)# Customer returns; warranty claims or service reports;

(n)# Review of staff resources, capability and training needs, further recruitment or reductions; performance of new starters; effectiveness of training recently provided;

(o)# Review of equipment performance, need for new, up-grading or maintenance;

(p)# Analysis and re-evaluation of suppliers/subcontractors and goods-inward inspection;

(q)# Review of corrective/preventive actions; also any potential dangers or adverse consequences on systems, controls, products, services or purchasing;

(r)# Review of Quality Objectives and Continual Improvement;

(s)# Review of effectiveness of QA system, any adverse or good trends with our product/ service or the QA system, with actions for improvements and further planning for Quality.

(t)# Review Quality Policy, is it still appropriate.

(*Author's note: the ones marked # are mandatory to meet the requirements of ISO 9000. The others are examples of items that are highly desirable to make this a sensible, interesting and relevant review. The order of each item is of no importance.*)

3.2 Implementation of Internal Audits

. . . . plc carries out internal audits to ensure our management systems are fully implemented and they provide effective controls for Quality Assurance. It also provides the Managing Director with an independent 'snapshot' of the whole company. The person doing the audit must have had basic training on 'ISO 9000 Internal Auditing' (**OR** have attended the . . . . course on Internal Auditing), and must be independent of the functions being audited, (**OR** this is a very small company of only 2 [**OR** 3 or 4] people, therefore it is not feasible to get complete independence. However, it has been arranged that this year one Director will audit the even number procedures and another the odd numbered procedures. At the end of the year the procedures being audited will reverse. Whilst this is not completely independent, it can be seen from the audit reports that the internal audits are thorough and effective with items being raised and corrected).

The policy is that each and every . . . . plc procedure will be audited at least once (**OR** twice) per year. However, as part of the ISO 9000 introductory programme there was an intensive round of audit cycle with a complete set of audits carried out for 3 or 4 months before the formal assessment.

No later than February of each year, the Quality Manager will draw up a schedule of internal audits, see Annex A. These are planned against each . . . . plc procedure in turn with a quarter of the procedures being done each three months (**OR** Procedure No. . . . . or Department . . . . is considered critical to the operations of . . . . plc. Hence they have been scheduled at . . . . . . number of audits per year).

The audit schedule includes a review of the Quality Manual and Procedures & DQS overall to ensure it still complies with ISO 9001:2000. (This can easily be checked by reviewing any changes made to the documented system and/or amendments made to ISO 9001 itself). Following the additional requirements of ISO 9001:2000 an audit is required of marketing material as they communicate marketing information, which must be correct.

The audits for 'Management Review' and 'Training' procedures will require careful selection of staff of appropriate level due to the confidentiality of these records. The nominated auditor arranges the audit date with the Departmental Head approximately two weeks prior to the audit. The Departmental Head will act as guide or nominate a deputy to remain with the Internal Auditor at all times. It is usually not necessary to give the auditor a copy of the previous audit report as it will have been closed out by the Quality Manager. For each audit there is an Audit Summary Report completed, see Annex B, and the audit is shown as completed on the annual plan. Any non-compliances are recorded, preferably at the time witnessed, on the Non-Compliance/Observation form, see Annex C. (**OR** to give an overall picture of company performance the non-compliance/observation arising from internal audits are recorded on the 'Poor Service Corrective Action Request Form', see Annex E.) Non-compliances will be raised when there is verifiable factual evidence seen or it is found that:

(a) Contract requirements are not being met;

(b) . . . . plc own specifications are not being met;

(c) The authorised procedures and instructions are not being complied with, or they are not available.

Observations will be noted if the internal auditor wishes to raise or record their personal concerns or suggestions for improvement. In addition the Internal Auditor may raise an observation where they believe the Quality Manual/Procedures/Instructions may give rise to non-compliances in the future. The forms also allow for records of the actions required. They must be completed within 2 months of raising or a shorter period if agreed and specified. The audit will then be shown as closed out on the annual internal audit plan. If judged appropriate, the Quality Manager may issue an instruction for a special re-audit if the results give concern. (**OR** for control purposes the Quality Manager maintains a log of Corrective Action Requests (see Annex F).)

The Quality Manager will write a short analysis and summary of the audit reports for submission to each management review (**OR** the copies of the reports are all stapled together and submitted for discussion at the Management Review Meeting).

Note: If a particular area gives poor results, the Quality Manager or Managing Director may decide, if appropriate, to submit the area to special re-audit in 3 months.

If appropriate, the Managing Director or Quality Manager may 'subcontract' the internal audits to an external body or individual with the appropriate training and experience. (**OR** . . . . plc is a sole practitioner. The internal audits have been formally subcontracted to . . . . who will carry out audits to an agreed schedule. The approved suppliers records show . . . .'s training record for carrying out audits). This may give the benefit of a completely independent outsider's view and may also provide very experienced assessors. The Managing Director or Quality Manager may authorise a special audit against a procedure, department or product, if appropriate.

3.3A Implementation corrective and preventive actions (typical production, engineering shop)

As items are found to be suspect they may be placed on temporary 'hold'. (See Operating Procedure 4 for description and controls for 'hold', 'quarantined', 'reject', etc.). This may be at goods-in, during manufacture or final test/despatch. The same controls will be given to items that have been returned to the factory generally known as 'customer returns'. If the inspector or supervisor/manager confirms they are defective, i.e. not to specification, they will be placed under 'quarantine' and, if physically possible, placed into the designated quarantine area (**OR** box, **OR** shelf **OR** . . . .). These defective items, or batches, are then reviewed with the following corrective actions considered:

(i)   Carry out additional work (e.g. rework) that will bring the items exactly to conforming with specification/drawing. These will be re-inspected to ensure they now conform to specification.

(ii)   'Sift' the items, where only a percentage of the items are defective, by 100% inspection, to remove the offending items.

(iii) Take managerial actions to accept the items 'as is' for their intended purpose, with the defects remaining by obtaining customer's formal written agreement to accept (**OR** raise a formal 'concession' or 'waiver' [see Annex H and J and entering onto the concession register (see Annex G) {**OR** book}]. A **minor concession** is one that does **not** affect functioning, safety, inter-changeability or life span/reliability and can be approved by the Quality Manager or the Managing Director. If the Quality Manager/ Director feels concern over a design aspect, they will request the advice and signature and approval of the Design Manager. Similarly if the Quality Manager/Director express doubt whether the minor defect will be accepted by the customer (e.g. visual appearance, different packaging etc.) they will seek the advice and signature of approval of the Sales/Marketing Director. If the Quality Manager/Design Manager or

<table>
<tr><td>Operating Procedure 7:   Issue . . . . . .</td></tr>
<tr><td>Dated . . . . / . . . . / 200 . . . .</td></tr>
<tr><td>Page 5 of 7</td></tr>
</table>

Sales/Marketing Director refuse to sign, it must be submitted as a customer concession or withdrawn. If the defect affects either functioning, safety, interchangeability or life span/reliability or it obviously does not conform to the accepted .... plc's standard, a customer (**OR** major) concession will be raised if appropriate).

(iv) Carry out remedial work that will make the items usable but not exactly to specification (e.g. repair, such as an unspecified weld repair on a crack in a shaft), hence, a concession or formal approval for this additional feature is required. These will be re-inspected to confirm that they conform to the agreed revised specification.

(v) Re-classify the non-conforming items to a lower grade of product (**OR** 'seconds').

(vi) Place under Red Card (see Operating Procedure 4) and scrap the items or returned purchased goods/material to the suppliers and make new, re-purchase or replace from stock.

(vii) Write to the customer advising of any defects in material or item that has been supplied as 'free issue'.

(ix) Also at this time it should be considered if there is any 'knock on' effect from this non-conformance and any actions required
e.g. Defective parts in stock
Defective parts already delivered and in circulation or use

The short-term corrective actions are recorded and authorised on the Product Corrective/Preventive Action Form shown at Annex D. The actions will also be authorised on the same form. The authorisation will be that of the Production Manager or Supervisor (**OR** the Director in their absence); who will also detail the loss in money or hours. The Production Manager and supervisor may ask the Quality Manager for advice on appropriate actions or the details of the fault, but the decision on the appropriate actions is theirs. A copy will be passed to the appropriate member of staff or operative to action, and passed to the Quality Manager for check and review when the work is completed, the original is filed in the workshop office (**OR** ....). After the short-term corrective action has been completed the Quality Manager will review the Corrective Action Form and any other information and, where appropriate, recommend long-term action to be taken. These will be considered at the Management Review.

The above is the procedure for all defective product arising in-house or where it is returned by the customer. Also of importance is 'Customer Complaints' that are received by letter, fax, phone or representatives reports. Every one of these will be allocated a reference of year/sequential number and entered into the Customer Complaint Record Book (**OR** form). Also returned product will be entered into the Complaints Record Book. The Complaints Record Book is headed year/number/date received/recorded by/client/description/date closed out.

In each case an acknowledgement letter summarising the complaint and promising investigation will be sent to the customer, preferably within three working days (**OR** .... days). The complaint will be thoroughly investigated and suitable actions proposed and taken. These will be recorded on the reverse side of the file copy of the acknowledgement letter sent to the client (**OR** on investigation form ....). When the investigation is complete, preferably within four weeks (**OR** .... weeks) an appropriate letter of apology/explanation will be sent to the client.

All the Corrective/Preventive Action Forms (**OR** and the approximate cost in lost material and man hours) and the entries in the complaints book will be considered at the Management Review meeting. In particular they will be reviewed for common features and adverse trends, etc.

(**OR** the majority of customer complaints, returns or guarantee claims are carried out by approved agents. They have prescribed proformas, see Annex . . . ., to complete to obtain payment of work/or replacement. These will be analysed and reported on to the management review meeting.)

(**OR** As part of our preventive action programme a number of products are purchased back after . . . . . . number of months [**OR** miles, etc.] and subject to a performance test in the . . . . and then a strip down critical inspection and formal report.)

3.3B Implementation corrective and preventive actions (typical service industry)

All cases of customer complaints or where it is found that service is not to the required specification will be recorded on the Poor Service Corrective Action Request Form (see Annex E) which are identified and filed in serial number order. The actions will be taken by the Quality Manager, who will be given support by other Managers/Directors as required.

This will also include non-compliances raised by external auditors from either the certification body or from clients, (**OR** also those recorded and actioned in our own internal audit systems/proformas will be recorded and actioned on the same Poor Service Corrective Action Request Form). These deficiencies in service will be reported together with the actions taken (in summarised version, or report showing numbers, trends, graphs, etc, if appropriate) to the Management Review meeting for their consideration. (**OR** . . . . plc are in a market sector by the very fact of providing a service to alleviate peoples' problems, we have innumerable 'whinges' that cannot be classed as complaints. Therefore by definition a Customer Complaint is one that is either a formal written complaint or a verbal complaint that the Managing Director (**OR** . . . .) considers so important that it should be formally recorded.)

(*Author's note: typical examples where it may be advisable to define 'complaints' as only those formally written are: 'meals on wheels to pensioners', 'nursing elderly patients', 'grass cutting gardens and maintenance on housing estates', 'maintaining photocopiers', 'vehicle recovery', 'courier services'.*)

Operating Procedure 7:	Issue . . . . . .
	Dated . . . . / . . . . / 200 . . . .
	Page 7 of 7

INTERNAL AUDITS PLANNED FOR 200. . . .

Quarter			J.F.M.	A.M.J.	J.A.S.	O.N.D.
Proc. No.	Auditor		Enter actual dates below			
1		Planned Completed Closed				
2		Planned Completed Closed				
3		Planned Completed Closed				
4		Planned Completed Closed				
5		Planned Completed Closed				
6		Planned Completed Closed				
7 Including review of DQS, Quality and Objectives Policy		Planned Completed Closed				
8		Planned Completed Closed				
9, 10, 11, etc.		Planned Completed Closed				
(**OR** Marketing Material)		Planned Completed Closed				
(**OR** Validation or Inspection)		Planned Completed Closed				

Annex A Oper. Proc. 7: Issue . . . . . .

Dated . . . . / . . . . / 200 . . . .

Page 1 of 1

AUDIT SUMMARY

Date: . . . .   Auditor: . . . .

Procedure/Department/Equipment/Product subject of Audit:

. . . .

Reasons for Audit standard/special: . . . .

References of documents checked, contract reference, training records, drawing numbers, etc.

. . . .

. . . .

. . . .

. . . .

Objective evidence recorded i.e. number of non-compliances, observations or defects:

. . . .

. . . .

Auditor's subjective comments or recommendations if appropriate:

. . . .

. . . .

Agreed actions/recommendations in addition to items on N/C-Obs. forms:

. . . .

. . . .

. . . .

Signed: . . . .   Signed: . . . .   Date: . . . .
               Auditor   Dept. Head

Annex B Oper. Proc. 7:   Issue . . . . . .

Dated . . . . / . . . . / 200 . . . .

Page 1 of 1

NON COMPLIANCE/OBSERVATION FORM

Ref No . . . .

Audit of: . . . .
(Procedure, Process, Department or Product, Supplier)

Factual evidence/or omission recorded/observations:

. . . .

. . . .

Non-compliance (or observations) against the requirements of:

. . . .

. . . .

. . . .

Agreed Corrective Actions:

. . . .

. . . .

. . . .

To be completed in 2 months, or . . . . days/weeks

Signed: . . . .   Signed: . . . .   Date: . . . .
            Auditor   Dept. Head

Confirmed as Actioned and Closed out:

. . . .

. . . .

. . . .

Signed: . . . .   Date: . . . .
            Auditor (**OR** Quality Manager)

Annex C Oper. Proc. 7:   Issue . . . . . .

Dated . . . . / . . . . / 200 . . . .

Page 1 of 1

PRODUCT CORRECTIVE/PREVENTIVE ACTION FORM

Serial No. _____

Item description: . . . .

Drawing or Part No: . . . .   Quantity: . . . . Estimate

Quantity: . . . . Confirmed

Attach inspectors report if applicable: . . . .

Where defect arose: . . . .

Sheet 1

Actions	Approx. cost material/man hours
1. Carry out additional work to bring to specification: Authorised . . . . date	
2. Write to customer to agree to accept 'as-is' (**OR** raise customer or minor concession) Authorised . . . . date	Reduction in price paid by customer? . . . . Estimated delay awaiting reply? . . . . days/weeks
3. Write to customer specifying the proposed repair and seeking agreement (**OR** raise customer or minor concession) Authorised . . . . date	Repair cost? . . . . Reduction in price? . . . . Possible delay . . . . days/weeks
4. Reduce items to lower grade as: . . . . Authorised . . . . date	Cost in reduction in worth? . . . .

Annex D Oper. Proc. 7:   Issue . . . . . .

Dated . . . . / . . . . / 200 . . . .

Page 1 of 2

Sheet 2

5. 100% inspect to remove defects	Cost to . . . . plc?
Inspection by	. . . .
Authorised . . . . date	Delays . . . . days/weeks
6. Scrap, or return to supplier, and replace	Cost to . . . . plc?
	. . . .
Authorised . . . . date	Delays . . . . days/weeks

Confirm items have been reworked/repaired, inspected again after completion:

Signed: . . . . dated . . . .

Confirmed signed . . . . dated . . . .
Quality Manager

Recommended long-term action to prevent recurrence in future:

Action by: . . . .

Signed: . . . . dated . . . .
(Quality Manager)

Circulate to: Management Review file

Production Manager, ... , ... , ...

Annex D Oper. Proc. 7:   Issue . . . . . .
Dated . . . . / . . . . / 200 . . . .
Page 2 of 2

POOR SERVICE   CORRECTIVE ACTION REQUEST   Serial No ....

Reported non-compliance, observation, customer complaint, or concern:

. . . .

. . . .

. . . .

From: . . . .   of . . . .

Taken by: . . . .   on . . . . date

Appropriate references: . . . .

. . . .

Short-term action:

. . . .

. . . .

. . . .

Signed: . . . .   Dated: . . . .

Confirm letter to client: Signed . . . .   Dated: . . . .

Review and proposed long-term corrective action:

. . . .

. . . .

. . . .

Signed: . . . .   Dated: . . . .

Remarks:

. . . .

. . . .

. . . . (Use overleaf if required)

Annex E Oper. Proc. 7:   Issue . . . . . .

Dated . . . . / . . . . / 200 . . . .

Page 1 of 1

LOG OF CORRECTIVE ACTION REQUESTS

Serial No.	Date raised	Location or customer	Brief description	Signed/dated Confirm closed out

Annex F Oper. Proc. 7:   Issue . . . . . .

Dated . . . . / . . . . / 200 . . . .

Page 1 of 1

CONCESSION REGISTER

Ref No. Year/No.	M or C	Date raised	Customer batch or project	Description

Annex G Oper. Proc. 7:   Issue . . . . . .

Dated . . . . / . . . . / 200 . . . .

Page 1 of 1

MINOR INTERNAL CONCESSION FORM ('M')

Concession reference No:
Customer/batch/project/order number:
Requested by:
Description of non-conformance and quantity involved:

Comments:	(i) Functioning affected YES/NO (ii) Safety affected YES/NO (iii) Interchangeable affected YES/NO (iv) Life span/reliability affected YES/NO

Agreement to concession (if appropriate/requested for advice)

Approved . . . .   Approved . . . .
    Signed: Director/Manager Design   Signed: Sales/Marketing Director/Manager

Approved Signature: . . . . (Quality Manager) Date:	Not Agreed Signature: . . . . Date: . . . .

Additional conditions of approval:

Signature: . . . .   Date: . . . .

CUSTOMER CONCESSION ('C') (**OR** MAJOR CONCESSION)

To: . . . .   Concession Ref No: . . . .

. . . .   Customer Ref/Order: . . . .

. . . . plc Ref/Order . . . .

Dear   ———————

We seek approval to make use of the following variations from your specifications/drawing/ contract requirement.

Note: Items marked # are to formalise agreed verbal or site instructions.

Signed . . . .   Date . . . .

Specifications/Drawings/ Contract Requirements	Proposed Variations	Reasons

The use of the above are approved on this contract:

Signed: . . . .   Position: . . . .   Date: . . . .

Please return to: . . . . plc, Tinatown, Blonkshire, in the enclosed stamped addressed envelope

Annex J Oper. Proc. 7: Issue . . . . . .

Dated . . . . / . . . . / 200 . . . .

Page 1 of 1

A draft model or workbook to prepare all your required Operating Procedures

.... plc
OPERATING PROCEDURE NO. 8

The procedure for control of documentation/data

Authorisation and Amendment Record

Issue No.	Date of issue	Prepared by	Authorised by:
1	Sept 20	A.N. Other	Draft for comment
2	Jan 20	A.N. Other	.... Signed: M. Director
3			
4			
5			
6			

Operating Procedure 8:   Issue ......

Dated ..../ ..../ 200....

Page 1 of 9

243

.... plc
OPERATING PROCEDURE NO. 8

The procedure for control of documentation/data

1. Purpose and scope

(a) Where .... plc has documents and data directly affecting the quality of our product and services, they must be controlled and of the correct issue. Such documents include:

   (i) contract requirements, specifications, drawings, quality or other plans, calculations, reports and contract amendments;

   (ii) internally generated, issued and authorised product or service specifications, drawings and instructions;

   (iii) International and national standards and specifications;

   (iv) the .... trade standards applicable to our industry as these are always (**OR** sometimes, **OR** occasionally) called up as a condition of our contracts (**OR** a requirement of our insurers, **OR** a legal requirement, **OR** a requirement to maintain our registration with the trade association ....);

   (v) the .... plc Quality Policy Manual and all Operating Procedures (**OR** and Standing and Temporary Work Instructions and Inspection Instructions);

   (**OR** .... plc are consultants, [**OR** recruiting agency **OR** ....] and customer letters and our replies and reports are an integral part of our 'product' or service. Hence all correspondence, whether externally or internally generated is formally controlled);

   (**OR** .... plc as an enterprise agency [**OR** Business Link, **OR** Chamber of Commerce **OR** Tourist information organisation **OR** ....] and are a provider of information. As these are an integral part of the service or product we supply they must be correct and of the appropriate issue);

   (**OR** .... plc are a planning control and approval authority, [**OR** photographers, **OR** model agencies, **OR** legal investigation **OR** ....], it is therefore vital that photographs are controlled to demonstrate what they are, when taken and by whom.)

(b) A member of .... plc when carrying out their duties must be working to and has ready access to, the appropriate issue of any documents. That withdrawn or obsolete issues are clearly identified as such, and are no longer readily available.

(c) Where documents are stored as electronic media these are also effectively controlled.

Operating Procedure 8:   Issue ......

Dated .... / .... / 200 ....

Page 2 of 9

2. Responsibility

The Director or Top Management is responsible for ensuring all Documents and data are of the correct issue. The Quality Manager will control the day-to-day issuing and controlling the Quality Manual and Operating Procedures (**OR** Standing or Temporary Work Instructions), all directors, managers and staff and operatives are to ensure that other related documents are issued correctly and also for ensuring that documents are treated with respect and are handled and filed (**OR** stored) in an appropriate manner.

3. Implementation – Issue of Quality Policy Manual and Operating Procedures (OR and Standing or Temporary Work Instructions)

3.1 Each and every page of the Quality Policy Manual and Operating Procedures (**OR** Standing or Temporary Work Instructions) have a control as a footer (**OR** header) identifying what it is, date of issue and issue number (**OR** date of issue only, to control the issue). Each of these documents have controls that show who formally authorised their issue.

3.2 For simplicity the Quality Policy Manual and each Operating Procedure with all their associated annexes, (**OR** Standing or Temporary Work Instructions) are issued as a whole. If a page is changed the whole manual or the whole individual procedure or instruction with their annexes, is amended to the next issue, with appropriate date, re-authorisation and circulated.

3.3 The circulation list for the Quality Policy Manual and Operating Procedures is shown on the front page of the Quality Policy Manual (**OR**, on separate control sheet). Again for simplicity and to prevent confusion and error the whole set of Quality Policy Manual and Operating Procedures is issued, contained in one lever arch file, to each holder. Whilst not every member of staff or operatives are issued with a copy of the Quality Policy Manual and Operating Procedures, each and every employee has direct unrestricted access to a copy and is encouraged to read the procedures relevant to their job, regularly.

3.4 It is intended that all external copies of the Quality Policy Manual are uncontrolled and will state at copy number 'N/A' and will be endorsed as 'uncontrolled' on the front sheet. If the client insists it is a controlled copy and the Managing Director give their approval, an external controlled copy will be added to the Circulation List.

3.5 No Operating Procedures (**OR** any documented Work Instructions) will be issued outside . . . . plc without the formal written instruction of the Managing Director.

3.6 (**OR** 'Standing Work Instructions' are issued where there are routine jobs that need additional specific instructions to operate a complex machine (**OR** . . . .). These are in addition and supplement the Operating Procedure and the instruction given to satisfy a particular contract. These are only issued to the location carrying out the specific routine operation. These are issued only via a 'Master List of Standing Work Instructions' controlled by the Quality Manager. This is headed 'Serial No/description/location/date issue/date withdrawn'. A master copy of every Standing Work Instruction issued, including the ones withdrawn are kept in a lever arch file by the Quality Manager. For simplicity when a Standing Work Instruction is amended it is withdrawn and replaced

Operating Procedure 8:   Issue . . . . . .

Dated . . . . / . . . . / 200 . . . .

Page 3 of 9

with the latest date of issue.) (**OR** 'Temporary Work Instructions' are issued for a one-off job e.g. job for one contract, production run, or a limited time period. These are authorised by the Department Manager and approved by the Quality Manager. The QA Manager keeps a log of all Temporary Work Instructions, with a photocopy of each one raised in a lever arch file.) See Annexes to Operating Procedure 3 for example of work instructions.

3.6 As the number issued of Quality Policy Manual and Operating Procedures is small and the locations are all within walking distance, the Quality Manager will personally change each issue and explain the changes to the holder. Hence no transmittal note system to record document control is necessary. (**OR** to record the changes made to the Quality Policy Manual and Operating Procedures to each recipient or site, use is made of the transmittal form shown at Annex A. These are returned to and checked by, the Quality Manager after each issue to ensure the transmittal records have all been returned by using form Annex B.)

3.7 At each amended issue of the Quality Policy Manual or the Procedures, in order to show the changes, so they can be easily seen by the reader, the new issues will have a vertical line adjacent to the change, in the right hand column. (**OR** the changes will be typed in italics) (**OR** underlined with red pen or by use of a highlighter.) (**OR** start with a # and finish off with a #. Two # side by side indicates there has been a deletion.) The Quality Manager will retain one copy of the replaced issues. These will be crossed through on each page and endorsed 'withdrawn'. Stapled to these will be a copy of the new issue. These copies of the old and new will be placed in a lever arch filed as 'Records of Amendments' and will be carefully kept so that the assessor from the independent third party certification body, customer or other appropriate bodies, can easily see the changes, controls and improvements to . . . . plc systems. If the changes are thought to be significant, or could affect . . . . plc's scope of registration, our certification body URS (**OR** BSI, NQA, TUV LRQA, etc.) will be informed in writing. (**OR** to carry out the above controls during the development, documentation and implementation of . . . . plc systems would have led to confusion and unnecessary conflict and hard work to no purpose and would have been counter productive. Hence on . . . . date, which was three months [**OR** four months] before the formal assessment all the uncontrolled draft copies of the Quality Policy Manuals, Operating Procedures (**OR** Standing Work Instructions) were withdrawn. A formal re-issue was made at that time [**OR** also the log of Standing Work Instructions]. Formal document control has been enforced and maintained since that date.) (**OR** . . . . plc have a network computer system. The Quality Manual and all the Operating Procedures have been made available as a paperless system on every screen in a 'read only' format. Hence no hard paper copy is issued. The QA Manager is the only person with access to amend the Quality Manual and Procedures. To demonstrate and record control the Master Copy is a one copy only, hard paper copy run off onto . . . . coloured paper, with the front page of the Quality Manual and each procedure signed and dated by the Quality Manager. If the Quality Manager changes the Manual or a procedure, two hard paper copies of the whole Manual or individual procedures are run off. One is signed and dated and included onto the 'Master'. The other copy is stapled together with the removed old one with each identified as 'old' and 'amended' with the changes underlined or highlighted, and kept in a lever arch file entitled 'Record of Changes'.) (**OR** any member of . . . . plc are encouraged and may recommend a change to the DQS, it will be sent to the Quality Manager [**OR** . . . .] via the Departmental Manager on the form shown at Annex E.)

4A. Implementation – control of international, British, internal, ... plc or Trade Standards

4.1 A master list of all the above documents held by . . . . plc is retained by the QA Manager (**OR** the librarian, **OR** the design manager **OR** . . . .). This is headed: 'Description/date of issue or ref./issuing authority/issued to/date amended/date withdrawn':

If there is only one copy held under 'Issued to', write 'master'.

If there are more than two copies issued each additional copy will be entered again on the master list e.g. for 3 copies held, i.e. 1 master and 2 issued, there will be three separate entries.

To re-issue a British or other standard use is made of the Transmittal Record of QA Documents see Annex A to record issue and withdrawal.

4.2 Within the above are several British Standards. The Quality Manager (**OR** . . . .) will obtain a monthly copy of 'BSI News'. This will be reviewed against the copies held and if any British Standard held has been superseded a new copy will be obtained, and the old copy withdrawn. (**OR** . . . . plc have joined 'BSI Plus' service (**OR** ILI, Technical Update) and they advise us when or if standards have changed.) (**OR** . . . . plc are consultants [**OR** contractors **OR** . . . .] engineers and have to conform or be aware of endless numbers of specifications and legislations and have therefore purchased with a maintenance contract a Barbour Index [**OR** . . . .]. This is controlled by the Quality Manager [**OR** . . . .] and is accessed by . . . .).

4B. Implementation control of international, British or trade standards for small firms holding only a copy of ISO 9001:2000

The products/services offered by . . . . plc do not normally require reference to British, international or trade standards.

The only formal standard held is ISO 9001:2000. As we are formally registered to ISO 9001:2000 standard by an Accredited Certification body it can safely be assumed that our Certification body, URS (**OR** . . . .), will inform us of any changes to the standard. (**OR** our Certification body insists we review at least on an annual basis that the standard is still current. In January [**OR** . . . .] each year the Quality Manager [**OR** . . . .] will phone BSI-Standards, or the appropriate national standards authority (or access the list freely available on the Internet)to ensure it is current [**OR** consult the controlled copy held by . . . .]. The front of the ISO 9001 standard will be endorsed 'Checked Current' and be signed and dated.)

If a client requests an item made to a British, International, Trade or Ministry of Defence (**OR** . . . .) standard, ... plc will request the client to provide a copy, or otherwise will purchase a copy for that order. This copy will not be controlled, the project/order number will be written in ink on the front cover. After the order is complete it will be placed within the appropriate contract file.

Operating Procedure 8: Issue . . . . . .

Dated . . . . / . . . . / 200 . . . .

Page 5 of 9

5. Implementation – contract information, drawings and specifications

5.1 All post arriving at . . . . plc (other than unsolicited junk mail and brochures) will be stamped with the . . . . plc date stamp. Information stapled or bound together need only have the top sheet date stamped. The office manager (**OR** typists **OR** receptionist **OR** . . . .) will do this task while sorting into piles for different recipients.

(**OR** within . . . . plc the control and confidentiality of documents and letters are absolutely vital. It is essential to be able to demonstrate that specific items of mail have been received or have been despatched. Each item of correspondence in addition to being stamped, will be entered into a 'mail-in' log, with headings: Date/day serial no/sender/brief description. When the recipient reads the document they will line through the date stamp and initial and date. They will also briefly note any actions taken.)

(**OR** Similarly items going out will be entered into 'mail out' log headed: Date/day serial no/ to . . . . ./ brief description/project No. The day serial number will be referenced on the actual correspondence, together with the project number.)

5.2 It is essential that all contract information is controlled so that it is not lost or inadvertently placed and mixed into another contract's information. Each sheet of information (**OR** the top sheet when stapled or bound together) is then identified with the Enquiry/Contract or Project Reference Serial Number (See Operating Procedure 1). Appropriate internal documents will also be identified with the Enquiry/Contract Reference Serial Number.

5.3 When reviewed or actioned the information is returned to the enquiry files. The enquiry files are loose yellow A4 envelope files kept in hanging files identified by completed tags, within the 4-drawer cabinet in Administration Office (**OR** . . . .). These are retained for at least 6 months. They are then reviewed by the director (**OR** . . . .) and may be scrapped as appropriate.

5.4 When the enquiry has become a contract and has been formally accepted (**OR** issued with a . . . . plc Work Order) the information is transferred to a green, spring spiral wired, file and transferred to a different 4-drawer filing cabinet.

(*Author's note: obviously this is not a requirement that the files be 'spring spiral wired', but I strongly recommend them if you have 'busy' files.*)

Each live contract file will be identified on the front (**OR** stapled to inside the front cover on form . . . ., see Annex _) with reference number, client details, brief job description, (**OR** for minor jobs, i.e. less than £. . . . ., all the information is put into a 'Minor Jobs' lever arch file with dividers between each job, with the job number on the divider).

5.5 The method of transferring this information from these files to the shop floor (**OR** . . . . .) is described in the Process Control Procedure 3.

| Operating Procedure 8: Issue . . . . . . |
| Dated . . . . / . . . . / 200 . . . . |
| Page 6 of 9 |

5.6 . . . . plc contracts usually only have one or two drawings, if any, from clients. These are stamped, signed and dated by . . . . and the . . . . plc project/order number clearly written on it with permanent ink. As there is automatically a copy taken (**OR** the original is always given to the workplace in the information job pack) and given to the workshop there is no need for a transmittal record.

(**OR** to maintain a record of which drawings and specifications are sent to which site or engineer, use is made of the transmittal record shown at Annex A). When the contract is completed one copy of each drawing or specification, is retained in the contract file.

(**OR** . . . . plc processes necessitate the handling of many customer supplied drawings and sketches . However it is still essential for our ISO 9001 registration that these drawings are issued and controlled effectively. It is essential that these are controlled as to work and deliver goods/services to the correct issue is absolutely vital. Therefore within each green contract file there is a 'Contract Register of Specification and Drawings', see Annex C this must be neatly and legibly completed. Any Quality Plan will follow the same controls.)

(**OR** . . . . plc drawings and specifications have to be controlled, as . . . . plc is an ISO 9001 approved design authority. The controls on issuing drawings and specifications are included in our Design Procedures.)

5.7 A copy of all correspondence sent to clients will be held in the file. It may be a photocopy of the original with the sender's signature or an additional copy run off the printer; the additional copy should be also signed or initialled by the sender to give positive evidence of despatch.

(**OR** . . . . plc is not a design authority. However, it does produce sketches to aid production. These sketches are done on A4 sheets and include the following control information on the sketch:

Description: _____
Project Title: _____
Enquiry/Contract No: _____
Originator: _____
Checked by: _____
Date: _____

After use in the shop floor [**OR** . . . .] they are retained in the contract file.)

(**OR** . . . . plc produces photos as part of its process. Each photo is identified on the back as follows:

Description: _____
Enquiry/contract Ref: _____
Photographer: _____
Date/approx. time: _____
(unless automatically provided)

These are kept in a clear plastic envelope sleeve in the contract file.)

Operating Procedure 8:   Issue . . . . . .

Dated . . . . / . . . . / 200 . . . .

Page 7 of 9

6. Brochures, catalogues, data sheets or leaflets

These are retained in the library by . . . . and are for general information only and are uncontrolled. As a new one is received the old version will either be scrapped, or if it contains useful information it will be lined through on the front cover and endorsed 'Superseded' and initialled and dated.

(**OR** . . . . plc products contain an important element of item/subassemblies designed and made by others. As these products are improving all the time, it is necessary to be fully aware of the specification of current models, whilst being able to identify obsolete models for spares. Hence a sticker/stamp is placed on the front of each catalogue stating:

Date received . . . .

Date superseded . . . .

The catalogues are controlled with a record book with the headings: Supplier/Catalogue description/Date received/Date superseded. The superseded catalogues are placed in a locked cupboard with the key held by . . . . As these are invaluable and in many cases irreplaceable they must not be removed from the . . . . office.)

(**OR** . . . . plc is an Enterprise Agency [**OR** Business Link **OR** Information Agency **OR** . . . .] hence the brochures and leaflets it provides are part of its product or service. Hence the Quality Manager [**OR** . . . .] maintains a master list of brochures/leaflets, see Annex D. If an instruction is given by the issuing authority the obsolete brochures/leaflets will be withdrawn from display/circulation and returned to the authority as 'client-supplied material'. [**OR** if permission is given in writing/by fax, by the authority they may be destroyed]. If no instruction is given by the issuing authority a subjective decision may be necessary as to whether to exhaust the stock of obsolete leaflets/brochures or withdraw and replace with new. If the old stock is to remain, a memo signed and dated by . . . . will be issued, otherwise the old stock will always be replaced.)

7. Electronic Data

Where documents are stored on electronic media it is essential that the following controls apply:

(i) that all floppy disks are identified
(ii) that management, design or quality system control documents have a 'Master' floppy disk (**OR** tape streamer) and a separate back-up disk (**OR** tape)
(iii) that the process of controlling these disks and backing them up is controlled on a formally authorised 'Standing Work Instruction'.
    (**OR** it is so easy for staff to alter information held on computer. Hence all management and quality system controlled documents [**OR** drawings with CAD systems] will have a 'master' hard paper copy. This stamped 'Master' is initialled and dated on each page by the Quality Manager and is held by the Quality Manager [**OR** . . . .] under lock and key.)
(iv) that disks or tapes are stored in such a manner as to prevent damage, corruption, loss or deterioration.

Operating Procedure 8:   Issue . . . . . .

Dated . . . . / . . . . / 200 . . . .

Page 8 of 9

8. Uncontrolled documents

There are cases where documents, booklets, tables, charts, etc. are no longer current but it is desirable to keep them as general background information or for legal records. All such documents will be stamped as 'Uncontrolled for general information only'. If the specific information is required for a project/contract a current copy of the information will be obtained.

9. Issue of the Quality Policy

The Quality Policy is displayed at strategic locations throughout the factory. As the number is small these will be personally controlled and changed by the Quality Manager (**OR** . . . .).

10. Manuscript amendments

Manuscript amendments to contract information, drawings, specifications or job cards are to be discouraged. When these are necessary the original information is to be neatly lined through, the revised information detailed and will be signed by authorised person and dated.

Typists' white amendment paint (Registered Trade name in UK is Tipp-Ex) must not be used to modify any issued contract information, drawings, specification, job cards, procedures or instructions.

(**OR** Limited use of Tipp-Ex is permitted for minor amendments on ORIGINAL DRAWINGS or the 'MASTER DRAWINGS' held in the Design Office. However, all such amendments must be detailed in the amendment/update index and be signed and dated).

11. Documents or Data that are 'Free-issue' or customer property

It is also understood that customer property can include intellectual property such as designs, drawings, commercial-in-confidence information, etc. Such customer property will remain 'commercial-in-confidence', it will not be photocopied for supply to others, or used by . . . . plc on our own products.

If we have visitors to . . . . plc, due care will be taken to ensure that the visitors do not have sight of their competitors' specifications or designs.

<table>
<tr><td>Operating Procedure 8:   Issue . . . . . .</td></tr>
<tr><td>Dated . . . . / . . . . / 200 . . . .</td></tr>
<tr><td>Page 9 of 9</td></tr>
</table>

TRANSMITTAL RECORD
QA DOCUMENTATION

To: . . . .   Copy No: . . . .

Please find attached/enclosed

. . . .   Issue . . . .   Dated . . . .

. . . .   Issue . . . .   Dated . . . .

. . . .   Issue . . . .   Dated . . . .

. . . .   Issue . . . .   Dated . . . .

. . . .   Issue . . . .   Dated . . . .

The changes introduced are:

. . . .

. . . .

. . . .

. . . .

Signed: . . . .   Dated . . . .
                    Quality Manager

Returned to Quality Manager Mr./Mrs. . . . .

I confirm I have read, understood and explained the above changes to the appropriate staff/ operatives. The documents have been replaced and the old copies destroyed (**OR** are attached).

Signed: . . . .   Dated: . . . .

Copy No:. . . . . . .

Note: If this confirmation is not received within 3 weeks it will be repeated with copy to the holders manager. If not returned within another 2 weeks it will be repeated with copy to the Director.

(Authors note: the above note is usually required only in larger organisations. If you do not insert this device, of increasing exposure or pain, the Quality Manager will be forever chasing transmittal slips.)

Annex A Oper. Proc. 8:   Issue . . . . . .

Dated . . . . / . . . . / 200 . . . .

Page 1 of 1

CONTROL SHEET QA DOCUMENT TRANSMITTAL CONTROL

Date of amendment: . . . .

Copy No.	Held by	Date of transmittal	Date chased	Date chased	Date chased	Date of transmittal return

Annex B Oper. Proc. 8:   Issue . . . . . .

Dated . . . . / . . . . / 200 . . . .

Page 1 of 1

CONTRACT REGISTER OF CLIENT SPECIFICATION AND DRAWINGS

Client: . . . .   Project: . . . .   Enquiry/Contract Ref: . . . .

Drawing/Specification/Title	Originator or Issue Authority	Date/ Issue No.	Date received	Issue to	Date withdrawn/ replaced

Annex C Oper. Proc. 8:   Issue . . . . . .

Dated . . . . / . . . . / 200 . . . .

Page 1 of 1

MASTER LIST OF BROCHURES/LEAFLETS, ETC.

Description	Issuing Authority	Date/ Issue No.	Ref. No. (where given)	Date received	Date on display	Date replaced or superseded	Date withdrawn or stock exhausted

Annex D Oper. Proc. 8: Issue . . . . . .

Dated . . . . / . . . . / 200 . . . .

Page 1 of 1

REQUEST TO CHANGE DOCUMENT

To: Quality Manager

via: Departmental Manager/Director

May I suggest and recommend a change to the following:

Quality Policy Manual   Issue . . . .   Page(s) . . . .

Operating Procedure No.   Issue . . . .   Page(s) . . . .

(**OR** Standing Work Instruction No: . . . .   Issue . . . .)

Form No: . . . .   Issue . . . .

Other Specification/Drawing/Description . . . .

Suggested or recommended Change:

Name (BLOCK CAPITALS) . . . .   Signed . . . .   Date . . . .
(P.T.O. if required or attach additional sheets)

Departmental Head/Directors comments:

Signed . . . .   Date . . . .

Quality Manager's review/action:

Signed . . . .   Date . . . .

Annex E Oper. Proc. 8:   Issue . . . . . .

Dated . . . . / . . . . / 200 . . . .

Page 1 of 1

.... plc
OPERATING PROCEDURE NO. 9

The procedure for quality, inspection and test records

Authorisation and Amendment Record

Issue No.	Date of issue	Prepared by	Authorised by:
1	5 Nov 19	A.N. Other	Initial Draft
2	12 Dec 19	A.N. Other	. . . . Signed: M. Director
3			
4			
5			
6			

Operating Procedure 9:   Issue . . . . . .

Dated . . . . / . . . . / 200 . . . .

Page 1 of 2

.... plc
OPERATING PROCEDURE NO. 9

The procedure for quality, inspection and test records

1. Purpose and scope

The purpose and scope of this procedure is to:

(a) summarise the contents of the other procedures with regard to maintaining records of Quality Documents, inspection and Test Records. These are covered in the other operating procedures but this will provide a complete checklist. There are other records kept but these are not controlled and are for general-business information or administration;

(b) summarise the records retain to assist . . . . plc defend any product liability claim. These are retained for 15 years. (**OR** . . . . plc products do not carry a significant product liability risk therefore records are retained for at least 3 years, to satisfy the 3 year audit cycle of our certification body).

(**OR** . . . . plc products designs and erects civil engineering projects. Appropriate records will be kept for 50 years (**OR** ...? years) or longer if specified in the contract).

(**OR** . . . . plc staff use chemical sprays. Appropriate spraying records will be kept for 60 years (**OR** ...? years) or longer if specified in the contract or a legal requirement).

2. Responsibility

The Directors or Top management have the responsibility to ensure records are adequate and correct.

The day-to-day administration has been delegated to the Quality Manager. All directors, managers, staff and operatives have responsibility for their appropriate records within this procedure.

3. Implementation

The records listed in Annex A are taken and maintained and filed in order to prevent loss and allow easy retrieval. The retention time and methods of disposal are also shown. In general all records will be identifiable to the applicable product, service, process product, etc. They will identify who completed the record by signature (**OR** identifiable initials) with the applicable date e.g. they will record 'what' the document is, 'who' authorised it, and 'when' it was issued or amended. All records will be in permanent material and be clearly legible.

(**OR** . . . . plc is a clothing factory. All records in the production areas have to be in pencil, to prevent permanent damage to cloth and garments by ink).

Operating Procedure 9:   Issue . . . . . .

Dated . . . . / . . . . / 200 . . . .

Page 2 of 2

Description	When, where recorded	Length of storage and location	Disposal
Management review minutes	Taken at each meeting	Retained minimum of 5 years in Management Review file	Disposed by tearing and binning
Drawings, specifications & Quality Plan	Identified as developed, recorded as formally issued or amended	Retain for at least 15 years, in respective contract file (**OR** in Drawing Office filing system)	Dispose by tearing
Review of tender submission	Made every time . . . . plc make an offer	Unsuccessful retain at least 6 months in enquiry file. Successful, retain 15 years in Contract file	Disposed by tearing and binning
Contract review with the customer order documents	Made every time . . . . plc accept a contract	Retain in Contract files which are kept for a minimum of 15 years	Disposed by tearing and binning
Approved subcontractors or suppliers	Maintained and amended continuously	The old lists will be filed and kept for at least 15 years, in QA Manager's records	Disposed by tearing and binning
Purchase orders	Made for each purchase	Retain Purchase Order Book for at least 15 years. To be stored by	Disposed by tearing and binning

Continued

Goods-inward or verification records, with appropriate suppliers' material certificates, inspection & traceability records	Made for each delivery	Retain with supplier's delivery note for at least 15 years. To be retained and maintained by	Disposed by tearing and binning
Rejected 'free-issue' material	Memo or letter for any occurrences	Retain in Contract file for at least 15 years	Disposed by tearing and binning
. . . . plc traceability records	Take on manufacture and maintained to delivery	Retained in Contract file for at least 15 years	Disposed by tearing and binning
(OR stores or warehouse inspections)	(OR routine check as per Operating Procedures)	(OR retain 15 years in QA Manager's records)	(OR disposed of by shredding)
Process, Inspection or Test Records (e.g. production test results, inspection verification results, log book & formal record diaries. Authority and verification signatures on instructions, drawings, etc. Certificates of Conformity, etc.)	Taken as arise. All such records will have identity of 'inspector' by signature and date of inspection, process or test	Within various proformas and record log as detailed in the Operating Procedure. Maintained for at least 15 years. Process control records to be retained by . . . . Inspection and QA records to be maintained and retained by the QA Manager	Disposed by tearing and binning (OR as a recruiting agency OR medical centre OR . . . . . . , these records are highly confidential. They will either be shredded on site or incinerated by a trusted member of staff or sent to approved disposal agency/firm). (OR as a contractor of Bridges and Structures it is essential that these records are retained for 60 years

			[OR .....], hence these are archived to Standing Working Instruction ...... and stored in .... [OR by Datavault Ltd. or .... an ISO 9000 registered company]
Calibration records	Taken during each calibration check	Retain by QA Manager for at least 15 years	Bin
Non-conformities and corrective/preventive action (e.g. scrap, rework, concessions, customer complaints, returns, and actions taken)	Taken as arise	Retain by QA Managers for at least 15 years. Concessions and customer correspondence and complaints, etc. retain for 10 years for QA Manager	Shred or incinerate
Internal & External Quality Audits	Taken as arise and as corrective actions take place	Retain by QA Manager for at least 3 years (essential for the independent third-party certification body audits)	Shred or incinerate
Personnel & Training Records	Created on commencement of employment and maintained continuously	Retain by .... for at least 3 years after employment ceased	Shred or incinerate
Safety Records	Created using forms and logs provided as test or process card out	Retain for at least 15 years, by .... (OR COSHH records concerning, poison, chemicals, pesticides and weedkillers to be retained for at least 60 years)	Shred or incinerate

.... plc
OPERATING PROCEDURE NO. 10

The procedure covering Quality Objectives and Continual Improvement

Authorisation and Amendment Record

Issue No.	Date of issue	Prepared by	Authorised by:
1	1 Jan 20	A.N. Other	Draft for comment
2	1 Feb 20	A.N. Other	Informal Issue
3	1 April 20	A.N. Other	2nd Informal Issue
4	1 July 20	A.N. Other	. . . . Signed: M. Director

Operating Procedure 10: Issue . . . . . .

Dated . . . . / . . . . / 200 . . . .

Page 1 of 2

.... plc
OPERATING PROCEDURE NO. 10

The procedure covering Quality Objectives and Continual Improvement

1. Purpose and scope

The purpose of this procedure is to ensure that the Top Management sets Quality Objectives that are important, achievable and measurable.

That there is a practical and managed process of continual or step-by-step improvement.

2. Responsibility

To ensure it is taken seriously and achieve results, the process is controlled by the Managing Director.

3. Implementation

The need for setting Quality Objectives and achieving continual improvement is addressed in the Quality Policy Statement and also in appropriate parts of the Quality Manual.

This is pursued on two levels:

    (a)   Overall system and management processes or company objectives,
    (b)   practical small quality improvement projects or jobs.

    (a)   Overall System or company objectives
           These are discussed, set and recorded as an agenda item at the Management Review Meeting.

    (b)   Practical small quality improvement projects or jobs
           At the Management Review meeting or throughout the year suggestions will be made to the Quality Manager for a job that can improve the system, the processes, the facilities or the product.

These will be recorded on the Proposal for Quality Improvement Project Form (Annex 'A').

When approved the Quality Manager will record the Quality Improvement job in a Log Book (see Annex 'B').

A small project file will be started by the designated project manager to record the programme and steps taken in the project to achieve a successful outcome. Once successful outcome the file will be passed to the Quality Manager and filed for reference.

(Author's note: for these projects to work successfully it may be necessary to provide training in basic problem solving techniques, graphs, Pareto analysis, fishbone diagrams, brainstorming, etc.)

Operating Procedure 10:   Issue ......

Dated ..../ ..../ 200....

Page 2 of 2

PROPOSAL FOR
QUALITY IMPROVEMENT PROJECT

Proposed Project: . . . .

. . . .

With the aim or purpose of:

. . . .

. . . .

. . . .

Perceived benefits: . . . .

. . . .

. . . .

Signed . . . . dated . . . .

Reviewed by Quality Manager:

Comments: . . . .

. . . .

. . . .

Resources required: . . . .

Likely timescale: . . . .

Project Manager
Proposal: . . . .

Signed . . . . dated . . . .

Managing Director's comments:

. . . .

. . . .

. . . .

Approved   Not Approved

Signed . . . . Managing Director

Date: . . . .

Annex A Oper. Proc. 10: . . . . . .

Dated . . . . / . . . . / 200 . . . .

Page 1 of 1

LOG OF QUALITY IMPROVEMENT JOBS

Job No.	Job Description	Project Manager	Date Approved Started	Outcome	Date Completed or closed

Annex B Oper. Proc. 10: . . . . . .

Dated . . . . / . . . . / 200 . . . .

Page 1 of 1

.... plc
OPERATING PROCEDURE NO. 11

The procedure covering Internal and External Communication

Authorisation and Amendment Record

Issue No.	Date of issue	Prepared by	Authorised by:
1	1 Jan 20	A.N. Other	Draft for comment
2	1 Feb 20	A.N. Other	Informal Issue
3	1 April 20	A.N. Other	2nd Informal Issue
4	1 July 20	A.N. Other	. . . .
			Signed: M. Director

Operating Procedure 11:   Issue . . . . . .

Dated . . . . / . . . . / 200 . . . .

Page 1 of 3

.... plc
OPERATING PROCEDURE NO. 11

The procedure covering internal and external communication

1. Purpose and scope
To demonstrate and explain to external bodies how these are co-ordinated.

2. Responsibility
Managing Director to ensure the processes exist and to encourage a culture of openness, lack of blame, with spirit of continual improvement for mutual benefit.

3. Implementation

External communication

At the Contract Review stage the offer document to be sent or the acceptance/or order are reviewed. If appropriate, or if we are in any doubt as to the stated requirements we will always and immediately communicate with the customer to ensure we are both exactly on the same wavelength.

At . . . . plc we are aware that it is vital to strive for excellence and to be the best. However this can be worthless if our customers do not have a high opinion or perception of our ability, or the customer is not satisfied. We seek to find out and raise the customer's perception of our strengths and abilities by:

(**OR** Company Brochures and leaflets)
(**OR** Company Newsletter)
(**OR** Questionnaire see Annex . . . .)
(**OR**. . . . . . . plc employ a Professional Marketing Manager)
(**OR**. . . . . . . plc employ salesmen, it is part of their job description and responsibility to effectively communicate with the customer)
(**OR**. . . . . . . ?)

Internal communication

As well as general information about the company and its products it **must include information on the effectiveness of the Quality System**. Also to get across the **relevance and importance** of all the employees' contributions. This is achieved via Management Review meetings where the managers and departmental representatives are requested and cascade suitable information down.

Also they are requested to solicit improvements and views from the shopfloor and intermediate level to cascade opinions, fears, suggestions etc. upwards.

Operating Procedure 11: Issue . . . . . .

Dated . . . . / . . . . / 200 . . . .

Page 2 of 3

It is considered essential that this happens. There are some MD's who are 'Sunshine Directors', e.g. they only want to hear good news or what they want to hear. This is bad for business. The MD is anxious to encourage suggestions and advice from all his workforce. (**OR** Internal Monthly newsletter.)

(**OR** Weekly 'Toolbox talks' where the departmental supervisor advises his staff of the current state of the company, new business, changes, improvements, errors made etc. The information is prepared for Monday morning by the MD. It is presented at afternoon tea break. The supervisor is instructed to cascade any comments or information from these talks back upwards to the MD.)

(**OR** posted weekly notice of company statistics, of performance and other information.)

(**OR** the MD will make a point of at least once per week walking the shopfloor and talking to the appropriate member of staff. This may be just for a chat, but if a worker wants to make a meaningful point the MD will ensure he makes appropriate time available to listen.)

(**OR** . . . . plc re-issue every two or three years an employee's handbook on the various products. This is provided for general background information but also to stress the relevance and importance of all employees' contributions to satisfying customers and the future success and survival of the company.)

Operating Procedure 11: Issue . . . . . .

Dated . . . . / . . . . / 200 . . . .

Page 3 of 3

Appendix F

Guidelines for preparing the operating procedures for Design

A quick review of clause 7.3 Design and Development will reveal there are seven sub-clauses e.g.:

7.3.1   Design and development planning
7.3.2   Design and development inputs
7.3.3   Design and Development outputs
7.3.4   Design and Development review
7.3.5   Design and development verification
7.3.6   Design and development validation
7.3.7   Control of Design and Development changes.

This may, at first, appear quite daunting but please do not be intimidated. If you have drafted your Quality Policy Manual for all the other clauses, you should now be familiar with the routine. Similarly, go through each sub-clause of the standard and reword it to state your general policy of what you will do (not in detail how you will do it) to control each of these particular design elements within . . . . plc.

Considering them and drafting this part of your Quality Policy Manual clause by clause should give you little problem if you are in fact carrying out original and intrinsic design.

The items which may give you problems for the Quality Policy Manual are:

7.3.4 Design Review
It is now **mandatory** to hold and record the minutes/notes/actions of at least one design review in any design project. This may come as a culture shock to those design offices who have taken the unapproachable, 'we know best', or 'ivory tower' approach. Your design proposals **must** be put up for criticism to representatives of Manufacturing, Purchasing, Sales/Marketing (or the customer, if appropriate) i.e. Can we make it? Can we buy the material and components? Is it what the customer wants?

7.3.5 Design verification
It is very difficult to show effective verification with a one-man design house, such as an independent architect. The standard does not state that the design elements will be checked by somebody independent of that design element. However, it can be a struggle to demonstrate alternative calculations without another qualified designer. With the advent of CAD (Computer Aided Design) Systems it may be possible to use a computer programme to carry out an independent verification check.

7.3.6 Design validation

This is a new element brought in by the 1994 version. In terms of manufactured product it is easy to explain and achieve.

Whereas 'verification' is a check on the various outputs of element of a **design** (e.g. calculation or drawing check), 'validation' is a check on the finished **product**. Typical examples will be:

- Working fatigue or stress reversal test or high speed/high load running tests on the final product in a laboratory. These would need careful documentation and planning to demonstrate that they were truly representative.
- Working installation, or 'field', trials in demanding applications with observations, readings and records taken.
- Production and issue of a trial batch under controlled conditions.
- Trial launch of a service under controlled and monitored conditions for a limited geographical area and time.

However with some service professions it is quite difficult to specify a validation, for instance where your service is not tangible e.g. an engineering or planning consultancy where the output is a report. In the 1994 version it was specified that you **shall** provide validation, so you could not write it out. This gave rise to some very creative writing to address validation e.g. 'if the customer wishes or agrees, a draft copy of the . . . . plc report will be sent to the client to validate, prior to final printing and issue'.

This has been relaxed and made more practical in the 9001:2000 version, adding 'WHERE PRACTICABLE, validation shall be completed prior to delivery'.

Additional note on validation on consumer products

Anyone making consumer products should read the Consumer Protection legislation and the implications on product liability. If somebody sues you under this act and you haven't got your validation trial results/reports available, you had better try and settle out of court, because **you are completely defenceless**. Do note that in addition to the organisation being sued or fined, there is a potential heavy **personal** liability for a fine or imprisonment if your product becomes defective and injures a member of the public. Also the firm can be forced into a product withdrawal which may cause them considerable difficulties.

Writing the operating procedures

Unfortunately the sub-clause headings shown in the ISO 9001 standard do not provide much guidance to help you to draft your operating procedures for design.

Such design procedures vary considerably and, as before, it is for you to document what you do and how you run your design office function, whilst meeting the words of each subclause of ISO 9001 and also your contract obligations.

However, I would advise that the design function usually drops conveniently into one of two situations and it is better to document and implement them differently. The two situations are:

- Design done continuously as an everyday event.
- Designs are developed irregularly, perhaps, once every 3 or 4 months.

I recommend that these two situations are approached completely differently.

Design as a continuous everyday activity

The design process should follow the logical steps shown in the diagram shown at the end of this appendix.

You may wish to reproduce this sketch and amend to exactly what you do and make it an annex to your design procedure. Having done that it is a relatively simple job to document the appropriate 'words' and produce a paragraph for each of the various steps to show how you plan and control design.

Design carried out irregularly

In my experience a procedure as above does not work if design is carried out infrequently. You simply forget what you did last time and have to re-discover the procedure each time, which is a problem.

In these cases I recommend a Design Project Quality Control Plan (DPQCP) that you complete each time as you do the design and you are then forced into the appropriate steps. This is completed and formally amended and re-issued as the job progresses to final issue. I attach a first draft of a DPQCP for you to amend and adapt, see following pages.

As the title 'Design Project Control Plan' is rather formidable, I recommend you change the title to something less frightening that suits your organisation e.g. 'Training Course Plan', 'New Product Development', etc.

Note: Conversely, if design is done continuously and is properly planned, controlled and recorded by your procedures a Design Project Quality Control Plan is probably not necessary (unless it is a specific contract requirement).

ISO 9001:2000 Quality Registration Step by Step

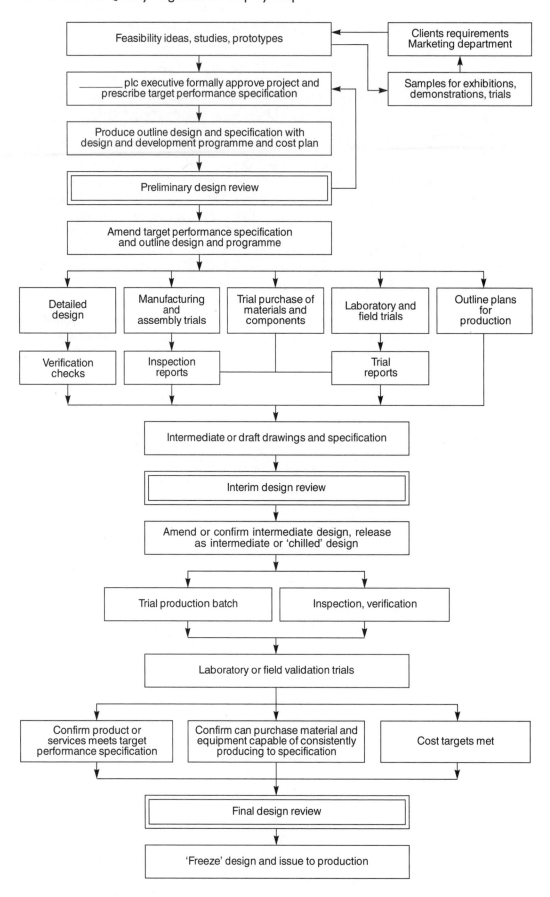

.... plc
DESIGN PROJECT
QUALITY CONTROL PLAN

(Sheet 1 of 3)

Project: ....

Reference: ....

Issue No.	Date	Authorised (Signed)	Circulated to
Draft			
1			
2			
3			
4			
5			
6			
7			
8			
9			
10			

DESIGN PROJECT, QUALITY CONTROL PLAN

(Sheet 2 of 3)

Project: . . . . Reference: . . . .

1. Project Manager: . . . .
   Others involved:
   . . . . Responsibilities . . . .
   . . . . Responsibilities . . . .
   . . . . Responsibilities . . . .

2. Arrangements for liaison and interfaces:
   . . . .
   . . . .
   . . . .

3. Target or required performance specification:
   . . . .
   . . . .
   . . . .
   . . . .

4. Details of project funding:
   . . . .
   . . . .
   . . . .

5. Target cost of final product design:
   . . . .
   . . . .

6. Any Statutory or Regulatory requirements:
   . . . .
   . . . .

7. Design input information:
   . . . .
   . . . .
   . . . .
   . . . .

8. Design output to be achieved and supplied, e.g. specification or drawings:
   . . . .
   . . . .
   . . . .
   . . . .

9. How are all the above to be reviewed, verified and approved (signed/dated) to show they meet the design input or contract requirement (any special acceptance criteria)?
   . . . .
   . . . .
   . . . .

10. Do the designs identify any features that are crucial for safety or proper functioning or reliability?

. . . .

. . . .

11. Design reviewed and agreed/approved by:

. . . . Position . . . . Date . . . .

. . . . Position . . . . Date . . . .

. . . . Position . . . . Date . . . .

12. Additional or special design reviews required:

. . . .

. . . .

. . . .

13. Arrangements to validate design e.g. working or field trials:

. . . .

. . . .

. . . .

14. Final design verification. All criteria met, design adequately specified. All amendments and modifications reviewed and incorporated.

Signed . . . . Dated . . . .

275

Appendix G

Notes of Process-based Quality Management System

The third edition of ISO 9001 (i.e. 9001:2000) in section 0.2 promotes the adoption of a 'process approach' when documenting and implementing an effective quality management system.

Note:

This does **not** mean that all processes have to be shown or documented as 'flow charts'. If you wish to or there is an advantage, or it aids clarity, to use flowcharts by all means do so.

Note:

This 'process approach' is nothing new or revolutionary. It has been promoted as a sensible, logical approach for many years.

The notes on 'process' in this book on pages 13 and 91 were in the original 1996 addition of this book.

Processes were explained in Frank Price's award winning book 'Right First Time' back in 1984 !

Some early documented systems, usually developed by consultants, in UK had a simplistic approach of a procedure for each of the 20 clauses of the 1979 or 1987 version of the standard. Therefore producing 20 Procedures:

e.g. Procedure 1 (4.1)   Management Responsibilities
      Procedure 2 (4.2)   Documented System
      Procedure 3 (4.3)   Contract Review
      Procedure 4 (4.4)   Design
      Procedure 5 (4.5)   Document Control
      Procedure 6 (4.6)   Purchasing
      etc., etc.
      Procedure 19 (4.19) Servicing
      Procedure 20 (4.20) Statistical Techniques.

Such a format satisfied the words of the standard, made it easy to document and audit. However it did not always or easily relate to the activities of many companies.

Also as some clauses did not in fact apply to a particular organisation some of these 'procedures' documented and presented simply stated that, 'this clause does not apply to . . . . .plc'.

Or worse still often these non-applicable clauses were filled with 'words' that meant absolutely nothing at all e.g. complete gibberish.

This simply confused the company and convinced many that these procedures and ISO 9000 registration was an essential, but totally useless, add-on that was required to get on tender lists and on contracts.

Over the years enlightened consultants have realised this and the majority of systems developed by UK consultants in recent years are already process-based. So instead of following the clauses of the standard, the procedures would follow the activities of the company. Also the number of procedures would be significantly reduced.

Typically existing process-based systems would have only 7–10 procedures

e.g.   Procedure 1   Sales/Marketing/Tender/Contract Review
         Procedure 2   Research, Design and Development
         Procedure 3   Suppliers, Purchasing, Goods-inward, Control of Materials
         Procedure 4   Manufacture, Test and Inspection, Packaging
         Procedure 5   Calibration
         Procedure 6   Training
         Procedure 7   Internal Audits, Complaints/feedback, Corrective/Preventive Actions, Management Review
         Procedure 8   Control of Documents and Data
         Procedure 9   Records.

Note:

When you first glance at the revised standard, it looks completely different to ISO 9001:1994

IN REALITY,

ITS EXACTLY THE SAME INGREDIENTS

IN A DIFFERENT ORDER WITH SOME MINOR

AMENDMENTS AND A FEW (BUT SIGNIFICANT) NEW BITS

Index

UNIVERSITY OF WALES, NE...
LIBRARY
AND
INFORMATION
SERVICES
ALLT-YR-YN

Index

Can you manage without it?

QA...

The Institute of Quality Assurance

Do you want to network with like-minded individuals to share best practice?

IQA Membership

Membership of the Institute of Quality Assurance (IQA) is an asset to any quality professional. IQA offers professional recognition, demonstrating an individual or company's commitment to quality. Members gain access to networking, *Quality World* magazine, preferential training rates, quality information and much more.

Do you need more information on quality?

Qualityworld

Qualityworld

Six sigma: it's coming for your organisation

the UK's leading quality magazine

IQA

With over 50 pages of editorial each month and all the latest jobs in quality, *Quality world* makes the ideal quality read. The magazine reports on the news, views and developments in quality management. It also covers contemporary business issues such as the environment, company culture, software and health and safety.

Subscribe today

For more information on the IQA and its products, or to subscribe to *Quality world*, please contact:
The Institute of Quality Assurance
12 Grosvenor Crescent, London SW1X 7EE
T: 020 7245 6722, F: 020 7245 6755
E: iqa@iqa.org,

IQA
Institute of Quality Assurance

www.iqa.org

Membership • Training • Education • Information • Publishing